No Metal No Magic

Element 95

Americium, Presented By Amerine

From The Magical Elements of the

Periodic Table Series

Amerine

Americium

By Sybrina Durant with Illustrations by Pranavva et al.

Americium, Presented By Amerine

From The Magical Elements of the

Periodic Table Series

Story copyright 2026

KDP Soft Cover ISBN: 9798258607843

BISAC Codes:

JNF051070 JUVENILE NONFICTION / Science & Nature / Chemistry

JNF016000 JUVENILE NONFICTION / Curiosities & Wonders

JNF051080 JUVENILE NONFICTION / Science & Nature / Earth Sciences / General

Soft Cover Print ISBN 13 - ISBN: 978-1-942740-65-0

Amerine The Actinide Knight Presents Americium

This Element 95 book features the periodic table element, Americium. It is presented by Amerine, a member of the Actinide Knights. Each knight has a magical sword or other medieval weapon tipped with an element that gives them unique powers. Their powers are based on the properties of its periodic table element.

Amerine is just one of the 118 elementals who will present all of the Magical Elements of the Periodic Table to readers who are curious about the wonders of the world.

Amerine introduces Americium in her book.

The Actinide Knights and their other techno-magical friends are the perfect group to introduce you to the elements in the Periodic Table. Hopefully, this Magical Elements of the Periodic Table book will spark an interest in the magical and real world properties of all the elements known today. You may be surprised at how prominently they feature in our every day lives.

Each page in this book contains terms that might not be completely familiar to the reader. Refer to the definitions in the back of the book to get a clear understanding of each meaning.

There is also a fun elemental themed Periodic Table at the back of the book. It features 118 elements presented by fanciful characters like unicorns, dragons, wizards, knights and goblins.. They want you to remember that if there's no metal...there's no magic or technology.

Remember, "No metal – No Magic. . .and No Technology".

It's Techo-Magical.

Note: Sybrina Publishing websites are Sybrina.com and MagicalPTElements.com. Follow sybrinapublishing on Instagram, Magical Elements of the Periodic Table on Facebook, @sybrinad on Pinterest, Sybrina_SPT on Twitter; and Sybrina Durant on LinkedIn.

Americium is an Actinide Metal

Americium (Am) was first produced in late 1944 by Glenn T. Seaborg, Ralph A. James, Leon O. Morgan, and Albert Ghiorso at the wartime Metallurgical Laboratory of the University of Chicago (now Argonne National Laboratory) in Illinois.

It is a synthetic, silvery-white radioactive metal that is both heat and electrically conductive.

It is paramagnetic, meaning it is weakly attracted to magnetic fields, over a wide temperature range, from liquid helium temperatures up to room temperature and above.

It is relatively soft, easily deformable, and more malleable than uranium or neptunium. It can be drawn into wires and pressed into thin sheets.

As a transuranium element, Americium is part of the radioactive, synthetic actinide row. Located below plutonium and curium, it falls within the Actinide series.

LEGEND

- Alkali Metals
- Alkali Earth Metals
- Transition Metals
- Post-Transition (or Other Metals)
- Metalloids
- Non-Metals
- Halogens
- Noble Gases
- Rare Earth Lanthanide Metals
- Actinide Metals
- Super Heavy—Radioactive

Americium Element

Atomic Structure

Actinide Metals—Any of a series of chemically similar metallic elements with atomic numbers ranging from 89 (actinium) to 103 (lawrencium). All of these elements are radioactive, and two of the elements, uranium and plutonium, are used to generate nuclear energy. The lanthanides and actinides are sometimes called the inner transition metals, referring to their properties and position on the table. They are actinium, thorium, protactinium, uranium, neptunium, plutonium, americium, curium, berkelium, californium, einsteinium, fermium, mendelevium, nobelium, and lawrencium.

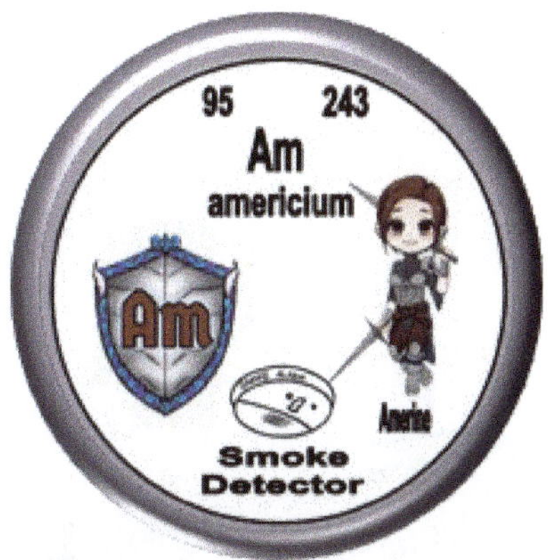

Americium is an intriguing element that belongs to the Actinide series in the periodic table. With the chemical symbol "Am" and an atomic number of 95, it is known for its radioactive properties. Discovered in the early 1940s, Americium is named after the Americas, reflecting its significance in American science and industry during that period.

Americium was first synthesized in 1944 by a team of scientists, including Glenn T. Seaborg, Albert Ghiorso, and Emilio Segrè. They discovered it through a process that involved bombarding plutonium with neutrons. This method helped create new elements, and Americium was one of the first synthetic elements to be produced. Due to its unique properties, it has garnered interest in various fields, particularly in nuclear science and technology.

One of the notable features of Americium is its radioactivity. Americium primarily emits alpha particles, which are a type of radiation made up of two protons and two neutrons. While alpha radiation is not highly penetrating and can be stopped by a sheet of paper or skin, it can be harmful if Americium is ingested or inhaled. Due to these properties, safety precautions are crucial when handling Americium, as exposure to radiation can lead to health issues, including cancer.

Americium is used in several practical applications. One of the most common uses is in smoke detectors. Americium-241, a specific isotope of Americium, is used in ionization smoke detectors. In these devices, the Americium emits alpha particles, which ionize the air in a small chamber. When smoke enters this chamber, it disrupts the flow of ions, triggering the alarm. This critical safety feature has made homes and buildings safer from fire hazards.

In addition to smoke detectors, Americium also has applications in industry and research. It can be found in certain types of gauges that measure thickness, density, or composition of materials. These gauges use Americium's radioactive properties to provide accurate measurements in manufacturing processes. Furthermore, Americium is utilized in some medical applications, particularly in radiation therapy and as a tracer in various diagnostic tests.

The production of Americium is usually done in nuclear reactors or through the reprocessing of spent nuclear fuel. Given its radioactive nature, handling Americium requires specialized facilities and trained personnel to ensure safety and minimize risks. Researchers continue to study Americium to understand its properties better and explore further potential applications in technology and medicine.

Despite its benefits, the use of Americium also raises environmental concerns. Radioactive waste management is a critical issue in nuclear science. Efforts are ongoing to find safe and effective ways to manage and dispose of radioactive materials, including those containing Americium. This is important to protect both human health and the environment.

Americium is a fascinating element with a blend of unique properties that make it valuable in various applications, particularly in safety devices and industrial measurements. While it offers significant benefits, its radioactive nature requires careful handling and consideration of safety protocols. As science advances, ongoing research may uncover even more ways to utilize Americium responsibly and effectively.

Uses For Americium

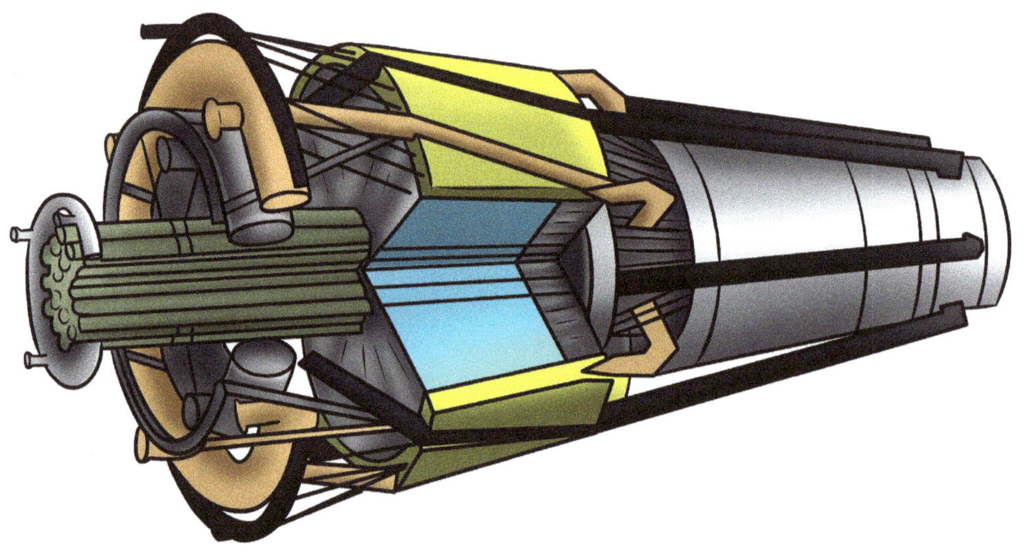

Americium, a synthetic element, has significant potential in the field of space exploration, particularly in its application for spacecraft nuclear batteries. These advanced energy sources are crucial for powering deep space missions, where traditional solar power becomes less effective due to distance from the Sun. By utilizing americium, spacecraft can benefit from longer-lasting and more reliable energy, thereby enhancing mission duration and enabling a greater range of scientific experiments far beyond our planet. This innovation represents a major advancement in space technology.

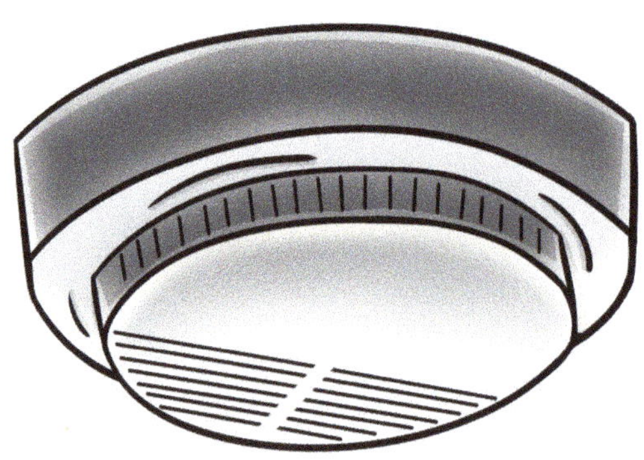

Americium plays a crucial role in enhancing the sensitivity of next-generation smoke detectors. This synthetic radioactive element, specifically Americium -241, is utilized in ionization smoke detectors due to its ability to emit alpha particles. By increasing the ionization process within the detection chamber, Americium significantly boosts the ability of these devices to identify smoke particles more effectively. As a result, home and commercial safety can be elevated, offering faster and more reliable alerts in the event of fire hazards.

Uses For Americium

(Continued)

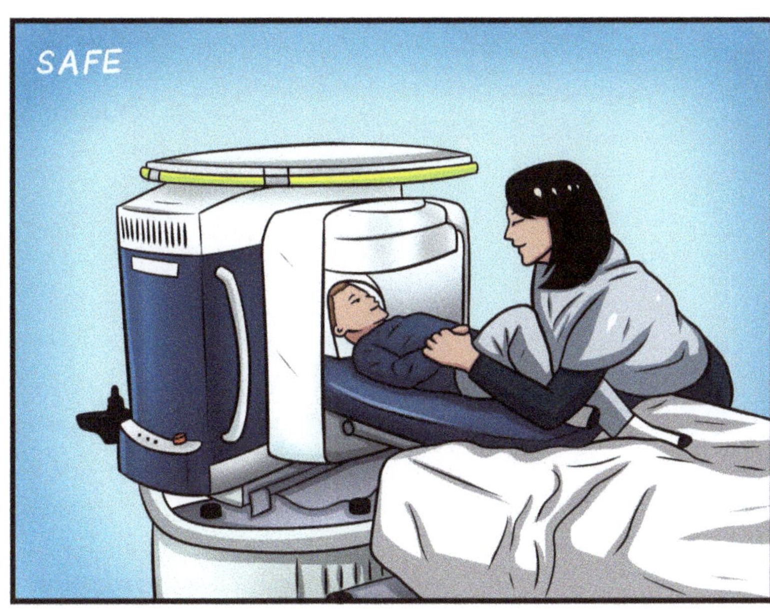

Americium plays a crucial role in portable medical imaging devices, particularly benefiting rural healthcare settings. Its use in these compact devices allows for high-quality imaging and accurate diagnoses, ensuring that patients in remote areas have access to essential medical services. By incorporating Americium, healthcare professionals can perform necessary examinations without the need for extensive medical facilities, ultimately improving patient outcomes. This advancement is vital for enhancing healthcare accessibility and ensuring timely medical interventions in underserved communities.

Smartphone batteries are now incorporating Americium, a radioactive element known for its ability to enhance energy storage capabilities. This innovation significantly extends battery life, providing users with longer-lasting performance between charges. As technology advances, the use of Americium in batteries could revolutionize power management in smartphones and other portable devices.

Uses For Americium

(Continued)

Air purification systems utilize Americium as a key component in their technology to effectively eliminate airborne pathogens and improve indoor air quality. This radioactive element is incorporated into specific filters or systems designed to capture and neutralize harmful microorganisms, such as bacteria, viruses, and other pollutants. By harnessing the unique properties of Americium, these purification systems enhance the overall safety and health of indoor environments, providing cleaner air for occupants and reducing the risk of respiratory issues and illnesses.

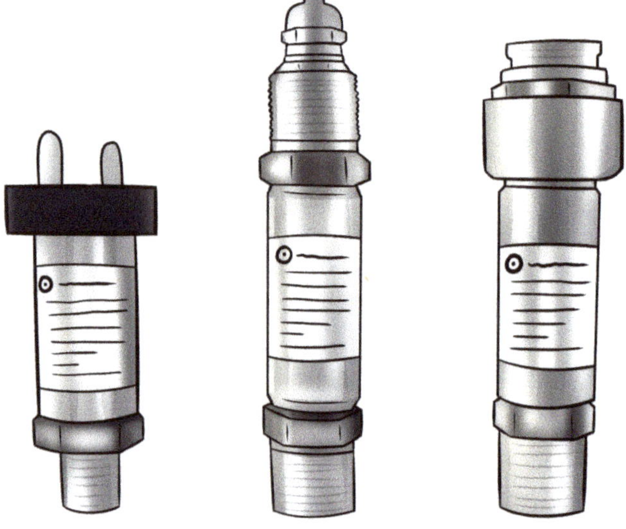

Industrial sensors are infused with Americium, a radioactive element that enhances their capabilities for high-precision manufacturing processes. By incorporating Americium into these sensors, manufacturers can achieve more accurate measurements of various parameters, leading to improved quality control and efficiency in production. This innovative use of Americium not only enhances the reliability of the data collected but also ensures that manufacturing standards are met with greater consistency, ultimately driving productivity and reducing waste in industrial operations.

The Source of Americium

Americium is a synthetic element that was first discovered in the 1940s. It is a man-made element, meaning it does not occur naturally in large amounts but is created through specific nuclear reactions. Americium is known for its radioactivity and has several important applications, most notably in smoke detectors and certain types of medical imaging. But how is this interesting element produced, and what is the story behind its main ingredient, plutonium-239?

Americium is produced by bombarding plutonium-239 with high-energy neutrons. This process primarily takes place in nuclear reactors, where the conditions are just right for such reactions to occur. Plutonium-239, an isotope of plutonium, plays a crucial role in this process. But first, let's understand what plutonium-239 is and why it was chosen for the production of americium.

Plutonium-239 is one of several isotopes of plutonium that can be used in nuclear reactions. It was selected for its capabilities during the mid-20th century when scientists were exploring ways to generate energy through fission reactions. Its favorable properties, like a relatively long half-life and the ability to readily absorb neutrons, made it the ideal candidate for further exploration.

The story of plutonium-239 begins during the Manhattan Project in World War II. Scientists were racing to develop nuclear weapons, and plutonium emerged as a valuable player in this quest. Plutonium was studied alongside uranium, with teams investigating different isotopes. The specific isotope, plutonium-239, was identified and isolated due to its ability to sustain a chain reaction—a vital characteristic for a nuclear weapon.

Once the war ended, the researchers were left with a deep understanding of plutonium-239. Its properties were further examined and harnessed for peaceful uses, such as in nuclear power plants. As the world began to recognize the potential dangers of radioactive materials, scientists sought ways to utilize nuclear processes for beneficial purposes. This shift led to the use of plutonium-239 not just for energy generation, but also for the production of other artificial elements, like americium.

The Source of Americium (continued)

To produce americium, plutonium-239 is placed in a nuclear reactor, where it is subjected to high-energy neutrons. When a neutron collides with a plutonium-239 nucleus, it can either cause the nucleus to split (fission) or be absorbed. In this case, the latter occurs—plutonium-239 absorbs the neutron, transforming into plutonium-240. Continuing this process, more neutrons can be absorbed, eventually yielding americium-241, a commonly used isotope of americium.

This process is often referred to as the "neutron bombardment" method. When plutonium-239 is bombarded, not only is americium produced, but a mixture of other isotopes can also result, including various isotopes of neptunium. The specific conditions in reactors, such as neutron flux and temperature, play a significant role in determining the efficiency of americium production.

Today, the production of americium involves more sophisticated and controlled methods than those first used during the early experiments. Nuclear reactors are operated under strict guidelines, and scientists continuously improve techniques to optimize the yield of americium-241.

One of the modern methods involves using fast reactors, where neutrons move at high speeds. These reactors create a suitable environment for producing americium efficiently. The americium is then separated from other elements through a series of chemical processes, ensuring a high purity level that is essential for its applications.

The extracted americium-241 can be in the form of pellets, fine powders, or other solid forms that is typically placed into small, sealed devices used in smoke detectors. The radioactivity of americium allows it to emit ions that can help detect smoke particles, making it a vital component in safety devices found in homes and businesses alike.

In summary, americium is an artificial element produced by bombarding plutonium-239 with high-energy neutrons in nuclear reactors. The journey of plutonium-239 started during World War II when it was identified for its unique properties that made it suitable for nuclear reactions. Today, advanced techniques and controlled environments are used to produce americium efficiently and safely, showcasing the ongoing innovations in nuclear science.

Understanding the complex processes that lead to the creation of americium can help us appreciate the intricate nature of nuclear chemistry and its importance in both historical contexts and modern applications. Whether it's through smoke detectors or other technological advancements, the legacy of americium continues to play a role in our daily lives.

Americium is located in Period 7 and is often considered part of Group 3 or is explicitly designated as an f-block actinoid.

The atomic symbol is Am. It's Atomic Number is 95. It's Atomic Mass is 243.00.

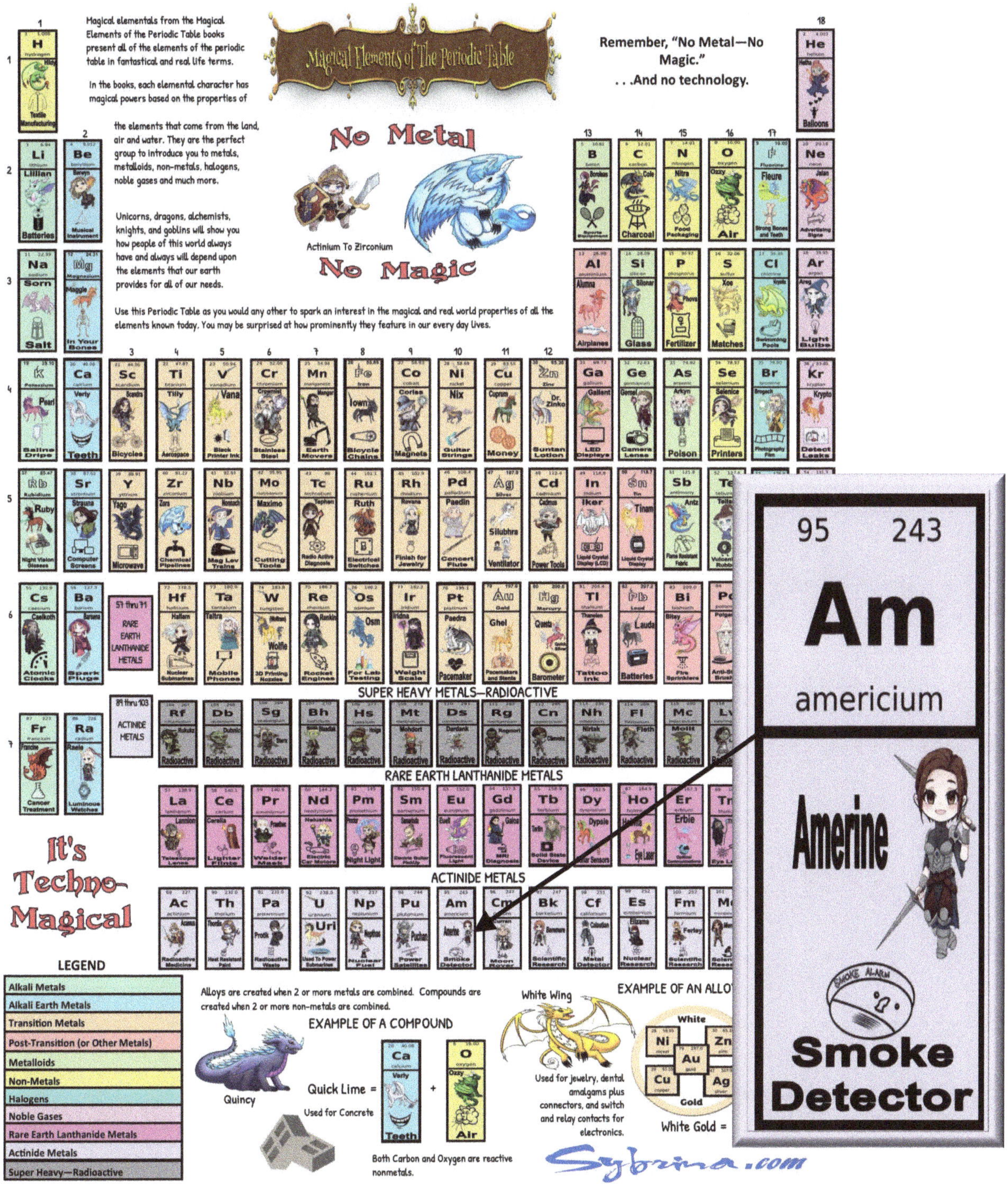

Magical Elements of The Periodic Table

Magical elementals from the Magical Elements of the Periodic Table books present all of the elements of the periodic table in fantastical and real life terms.

In the books, each elemental character has magical powers based on the properties of the elements that come from the land, air and water. They are the perfect group to introduce you to metals, metalloids, non-metals, halogens, noble gases and much more.

Unicorns, dragons, alchemists, knights, and goblins will show you how people of this world always have and always will depend upon the elements that our earth provides for all of our needs.

Use this Periodic Table as you would any other to spark an interest in the magical and real world properties of all the elements known today. You may be surprised at how prominently they feature in our every day lives.

No Metal
Actinium To Zirconium
No Magic

Remember, "No Metal— No Magic."
. . .And no technology.

H hydrogen 1 1.008 Hildy Textile Manufacturing
He helium 2 4.003 Hetha Balloons
Li lithium 3 6.94 Lillian Batteries
Be beryllium 4 9.012 Berwyn Musical Instrument
B boron 5 10.81 Boroleas Sports Equipment
C carbon 6 12.01 Cole Charcoal
N nitrogen 7 14.01 Nitra Food Packaging
O oxygen 8 16.00 Ozzy Air
F fluorine 9 19.00 Fleure Strong Bones and Teeth
Ne neon 10 20.18 Jalan Advertising Signs
Na sodium 11 22.99 Sorn Salt
Mg magnesium 12 24.31 Maggie In Your Bones
Al aluminium 13 26.98 Alumna Airplanes
Si silicon 14 28.09 Silonar Glass
P phosphorus 15 30.97 Phova Fertilizer
S sulfur 16 32.06 Xoe Matches
Cl chlorine 17 35.45 Krystix Swimming Pools
Ar argon 18 39.95 Areg Light Bulbs
K potassium 19 39.10 Pearl Saline Drips
Ca calcium 20 40.08 Verly Teeth
Sc scandium 21 44.96 Scandra Bicycles
Ti titanium 22 47.87 Tilly Aerospace
V vanadium 23 50.94 Vana Black Printer Ink
Cr chromium 24 52.00 Crowmist Stainless Steel
Mn manganese 25 54.94 Mangar Earth Movers
Fe iron 26 55.85 Iown Bicycle Chains
Co cobalt 27 58.93 Coriss Magnets
Ni nickel 28 58.69 Nix Guitar Strings
Cu copper 29 63.55 Cuprum Money
Zn zinc 30 65.38 Dr. Zinko Suntan Lotion
Ga gallium 31 69.72 Gallant LED Displays
Ge germanium 32 72.63 Gemel Camera Lense
As arsenic 33 74.92 Arkyn Poison
Se selenium 34 78.97 Selenice Printers
Br bromine 35 79.90 Brogach Photography Film
Kr krypton 36 83.80 Krypto Detect Leaks
Rb rubidium 37 85.47 Ruby Night Vision Glasses
Sr strontium 38 87.62 Strauna Computer Screens
Y yttrium 39 88.91 Yago Microwave
Zr zirconium 40 91.22 Zora Chemical Pipelines
Nb niobium 41 92.91 Nonnach Mag Lev Trains
Mo molybdenum 42 95.95 Maximo Cutting Tools
Tc technetium 43 98 Tephen Radio Active Diagnosis
Ru ruthenium 44 101.1 Ruth Electrical Switches
Rh rhodium 45 102.9 Rovana Finish for Jewelry
Pd palladium 46 106.4 Paedin Concert Flute
Ag silver 47 107.9 Silubhra Ventilator
Cd cadmium 48 112.4 Cadmus Power Tools
In indium 49 114.8 Iker Liquid Crystal Display (LCD)
Sn tin 50 118.7 Tinam Liquid Crystal Display
Sb antimony 51 121.8 Antz Flame Resistant Fabric
Te tellurium 52 127.6 Tellan Vulcanized Rubber
I iodine 53 126.9 Jody (Jodium) Cloud Seeding
Xe xenon 54 131.3 Xena Used To Catch Speeders

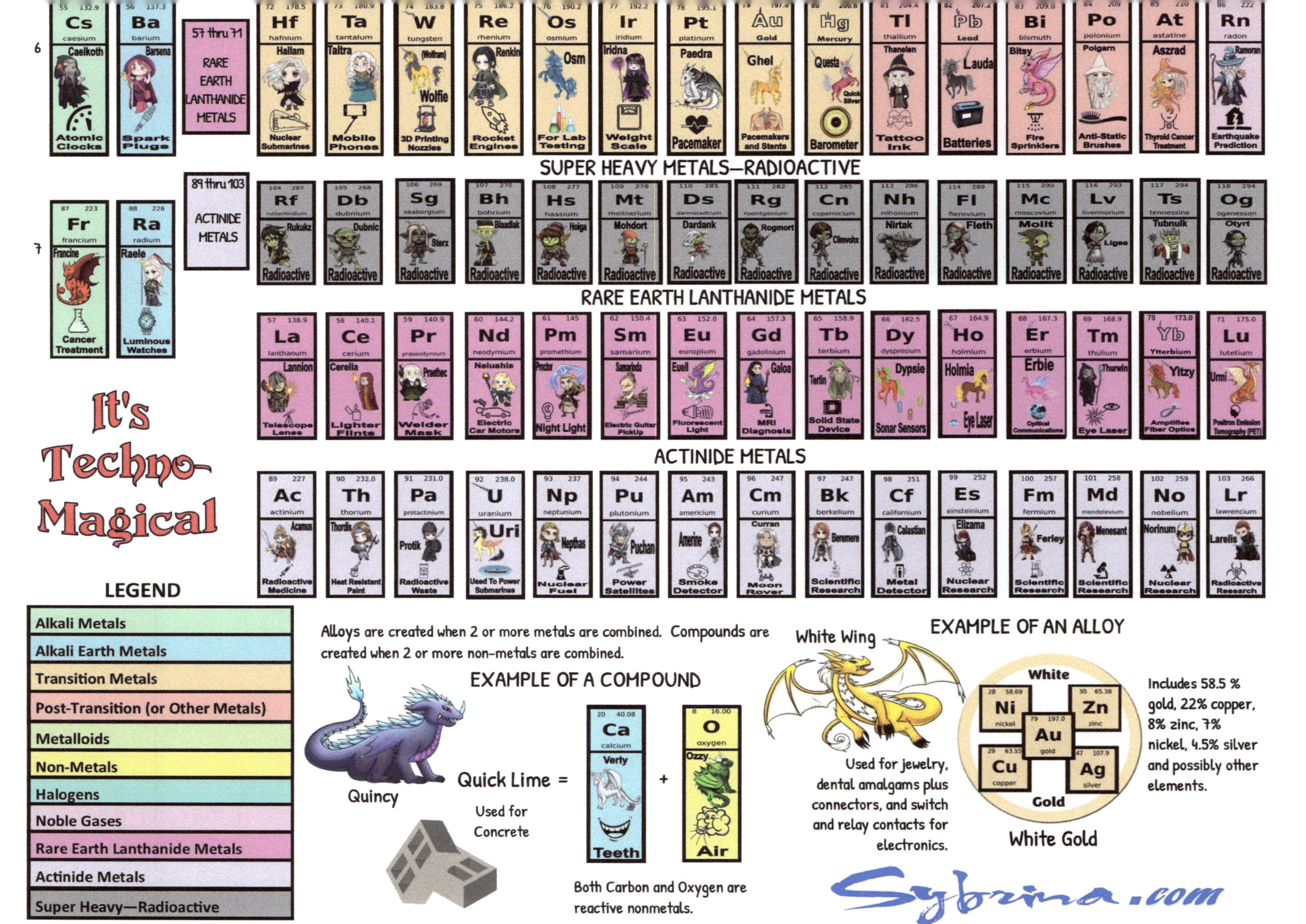

It's Techno-Magical

SUPER HEAVY METALS—RADIOACTIVE
RARE EARTH LANTHANIDE METALS
ACTINIDE METALS

Cs 55 132.9 caesium Caelkoth — Atomic Clocks
Ba 56 137.3 barium Barsena — Spark Plugs
57 thru 71 RARE EARTH LANTHANIDE METALS
Hf 72 178.5 hafnium Hallam — Nuclear Submarines
Ta 73 180.9 tantalum Taltra — Mobile Phones
W 74 183.8 tungsten (Wolfram) Wolfie — 3D Printing Nozzles
Re 75 186.2 rhenium Renkin — Rocket Engines
Os 76 190.2 osmium Osm — For Lab Testing
Ir 77 192.2 iridium Iridna — Weight Scale
Pt 78 195.1 platinum Paedra — Pacemaker
Au 79 197.0 Gold Ghel — Pacemakers and Stents
Hg 80 200.6 Mercury Questa Quick Silver — Barometer
Tl 81 204.4 thallium Thanelen — Tattoo Ink
Pb 82 207.2 Lead Lauda — Batteries
Bi 83 209.0 bismuth Bitsy — Fire Sprinklers
Po 84 209 polonium Polgarn — Anti-Static Brushes
At 85 210 astatine Aszrad — Thyroid Cancer Treatment
Rn 86 222 radon Ramoran — Earthquake Prediction

Fr 87 223 francium Francine — Cancer Treatment
Ra 88 226 radium Raele — Luminous Watches
89 thru 103 ACTINIDE METALS

Rf 104 267 rutherfordium Rukukz — Radioactive
Db 105 268 dubnium Dubnic — Radioactive
Sg 106 269 seaborgium Sterx — Radioactive
Bh 107 270 bohrium Blaadlak — Radioactive
Hs 108 277 hassium Hoiga — Radioactive
Mt 109 278 meitnerium Mohdort — Radioactive
Ds 110 281 darmstadtium Dardank — Radioactive
Rg 111 282 roentgenium Rogmort — Radioactive
Cn 112 285 copernicium Climvolix — Radioactive
Nh 113 286 nihonium Nirtak — Radioactive
Fl 114 289 flerovium Fleth — Radioactive
Mc 115 290 moscovium Molit — Radioactive
Lv 116 293 livermorium Ligee — Radioactive
Ts 117 294 tennessine Tubnulk — Radioactive
Og 118 294 oganesson Otyrt — Radioactive

La 57 138.9 lanthanum Lannion — Telescope Lense
Ce 58 140.1 cerium Cerelia — Lighter Flints
Pr 59 140.9 praseodymium Praethec — Welder Mask
Nd 60 144.2 neodymium Nelushla — Electric Car Motors
Pm 61 145 promethium Proctor — Night Light
Sm 62 150.4 samarium Samarinda — Electric Guitar PickUp
Eu 63 152.0 europium Euell — Fluorescent Light
Gd 64 157.3 gadolinium Galoa — MRI Diagnosis
Tb 65 158.9 terbium Terlin — Solid State Device
Dy 66 162.5 dysprosium Dypsie — Sonar Sensors
Ho 67 164.9 holmium Holmia — Eye Laser
Er 68 167.3 erbium Erbie — Optical Communications
Tm 69 168.9 thulium Thurwin — Eye Laser
Yb 70 173.0 Ytterbium Yitzy — Amplifies Fiber Optics
Lu 71 175.0 lutetium Urml — Positron Emission Tomography (PET)

Ac 89 227 actinium Acamus — Radioactive Medicine
Th 90 232.0 thorium Thordis — Heat Resistant Paint
Pa 91 231.0 protactinium Protik — Radioactive Waste
U 92 238.0 uranium Uri — Used To Power Submarines
Np 93 237 neptunium Nepthas — Nuclear Fuel
Pu 94 244 plutonium Puchan — Power Satellites
Am 95 243 americium Amerine — Smoke Detector
Cm 96 247 curium Curran — Moon Rover
Bk 97 247 berkelium Beremere — Scientific Research
Cf 98 251 californium Calastian — Metal Detector
Es 99 252 einsteinium Elizama — Nuclear Research
Fm 100 257 fermium Ferley — Scientific Research
Md 101 258 mendelevium Menesant — Scientific Research
No 102 259 nobelium Norinum — Nuclear Research
Lr 103 266 lawrencium Larelis — Radioactive Research

LEGEND
Alkali Metals
Alkali Earth Metals
Transition Metals
Post-Transition (or Other Metals)
Metalloids
Non-Metals
Halogens
Noble Gases
Rare Earth Lanthanide Metals
Actinide Metals
Super Heavy—Radioactive

Alloys are created when 2 or more metals are combined. Compounds are created when 2 or more non-metals are combined.

EXAMPLE OF A COMPOUND
Quincy
Quick Lime =
Ca 20 40.08 calcium Verty — Teeth
+
O 8 16.00 oxygen Ozzy — Air
Used for Concrete
Both Carbon and Oxygen are reactive nonmetals.

White Wing
Used for jewelry, dental amalgams plus connectors, and switch and relay contacts for electronics.

EXAMPLE OF AN ALLOY
White
Ni 28 58.69 nickel
Zn 30 65.38 zinc
Au 79 197.0 gold
Cu 29 63.55 copper
Ag 47 107.9 silver
Gold
White Gold
Includes 58.5 % gold, 22% copper, 8% zinc, 7% nickel, 4.5% silver and possibly other elements.

Sybrina.com

All Of The Periodic Table Elements Listed Alphabetically

ACTINIUM—*AC*—89

ALUMINUM—*AL*—13

AMERICIUM—*AM*—95

ANTIMONY—*SB*—51

ARGON—*AR*—18

ARSENIC—*AS*—33

ASTATINE—*AT*—85

BARIUM—*BA*—56

BERKELIUM—*BK*—97

BERYLLIUM—*BE*—4

BISMUTH—*BI*—83

BOHRIUM—*BH*—107

BORON—*B*—5

BROMINE—*BR*—35

CADMIUM—*CD*—48

CALCIUM (Vital)—*CA*—20

CALIFORNIUM—*CF*—98

CARBON—*C*—6

CERIUM—*CE*—58

CESIUM—*CS*—55

CHLORINE (Keen)—*CL*—17

CHROMIUM—*CR*—24

COBALT—*CO*—27

COPERNICIUM—*CN*—112

COPPER—*CU*—29

CURIUM—*CM*—96

DARMSTADTIUM—*DS*—110

DUBNIUM—*DB*—105

DYSPROSIUM—*DY*—66

ERBIUM—*ER*—68

EINSTEINIUM—*ES*—99

EUROPIUM—*EU*—63

FERMIUM—*FM*—100

FLEROVIUM—*FL*—114

FLUORINE—*F*—9

FRANCIUM—*FR*—87

GADOLINIUM—*GD*—64

GALLIUM—*GA*—31

GERMANIUM—*GE*—32

GOLD—*AU*—79

HAFNIUM—*HF*—72

HASSIUM—*HS*—108

HELIUM—*HE*—2

HOLMIUM—*HO*—67

HYDROGEN—*H*—1

INDIUM—*IN*—49

IODINE (JODIUM)—*I*—53

IRIDIUM—*IR*—77

IRON—*FE*—26

KRYPTON—*KR*—36

LANTHANUM—*LA*—57

LAWRENCIUM—*LR*—103

LEAD—*PB*—82

LITHIUM—*LI*—3

LIVERMORIUM—*LV*—116

LUTETIUM (Unique)—*LU*—71

MAGNESIUM—*MG*—12

MANGANESE—*MN*—25

MEITNERIUM—*MT*—109

MENDELEVIUM—*MD*—101

MERCURY (QUICK SILVER)—*HG*—80

MOLYBDENUM—*MO*—42

MOSCOVIUM—*MC*—115

NEODYMIUM—*ND*—60

NEON (Jazzy)—*NE*—10

NEPTUNIUM—*NP*—93

NICKEL—*NI*—28

NIHONIUM—*NH*—113

NIOBIUM—*NB*—41

NITROGEN—*N*—7

NOBELIUM—*NO*—102

OGANESSON—*OG*—118

OSMIUM—*OS*—76

OXYGEN—*O*—8

PALLADIUM—*PD*—46

PHOSPHORUS—*P*—15

PLATINUM—*PT*—78

PLUTONIUM—*PU*—94

POLONIUM—*PO*—84

POTASSIUM—*K*—19

PRASEODYMIUM—*PR*—59

PROMETHIUM—*PM*—61

PROTACTINIUM—*PA*—91

RADIUM—*RA*—88

RADON—*RN*—86

RHENIUM—*RE*—75

RHODIUM—*RH*—45

ROENTGENIUM—*RG*—111

RUBIDIUM—*RB*—37

RUTHENIUM—*RU*—44

RUTHERFORDIUM—*RF*—104

SAMARIUM—*SM*—62

SCANDIUM—*SC*—21

SEABORGIUM—*SG*—106

SELENIUM—*SE*—34

SILICON—*SI*—14

SILVER—*AG*—47

SODIUM—*NA*—11

STRONTIUM—*SR*—38

SULFUR (Xanthous)—*S*—16

TANTALUM—*TA*—73

TECHNETIUM—*TC*—43

TELLURIUM—*TE*—52

TENNESSINE—*TS*—117

TERBIUM—*TB*—65

THALLIUM—*TI*—81

THORIUM—*TH*—90

THULIUM—*TM*—69

TIN—*SN*—50

TITANIUM—*TI*—22

TUNGSTEN—*W* (WOLFRAM)—74

URANIUM—*U*—92

VANADIUM—*V*—23

XENON—*XE*—54

YTTERBIUM—*YB*—70

YTTRIUM—*Y*—39

ZINC—*ZN*—30

ZIRCONIUM—*ZR*—40

- Americium is named after the Americas in a nod to the naming pattern of the periodic table, since it sits directly below europium, which was named after Europe. This connection reflects a tradition in chemistry of honoring geographic regions and important cultural landmarks through element names. Americium itself is a synthetic element, first produced in a laboratory, and its name helps link modern scientific discovery with a broader historical context. The parallel between americium and europium makes the name especially memorable, showing how the periodic table often preserves subtle relationships between elements through both their properties and their names.

- Beginning in 2004, the United States ceased domestic production of Americium, creating a significant gap in the national supply chain for this important isotope. As a result, Am-241 was not available for commercial sale in the United States for many years. In 2020, the Department of Energy recognized the strategic and scientific importance of restoring a long-term domestic supply and took steps to reestablish access. This decision helped address critical needs in research, industry, and safety applications that depend on Am-241. The restoration of availability marked an important move toward strengthening U.S. isotope production and reducing reliance on foreign sources.

- It is the one actinide that is found in almost every household: as an isotope that produces ionizing radiation, it is an essential component of smoke detectors. The americium compound emits alpha particles that strike air molecules in their path, causing them to ionize.

Structure Of Elements In The Periodic Table

Periodic tables are laid out in rows and columns.

Vertical columns are called Groups.

Each element is placed in a specific location because of its atomic structure. Elements are arranged in Families.

Horizontal rows are called Periods.

All rows read left to right.

The last two periods are part of periods 6 & 7.

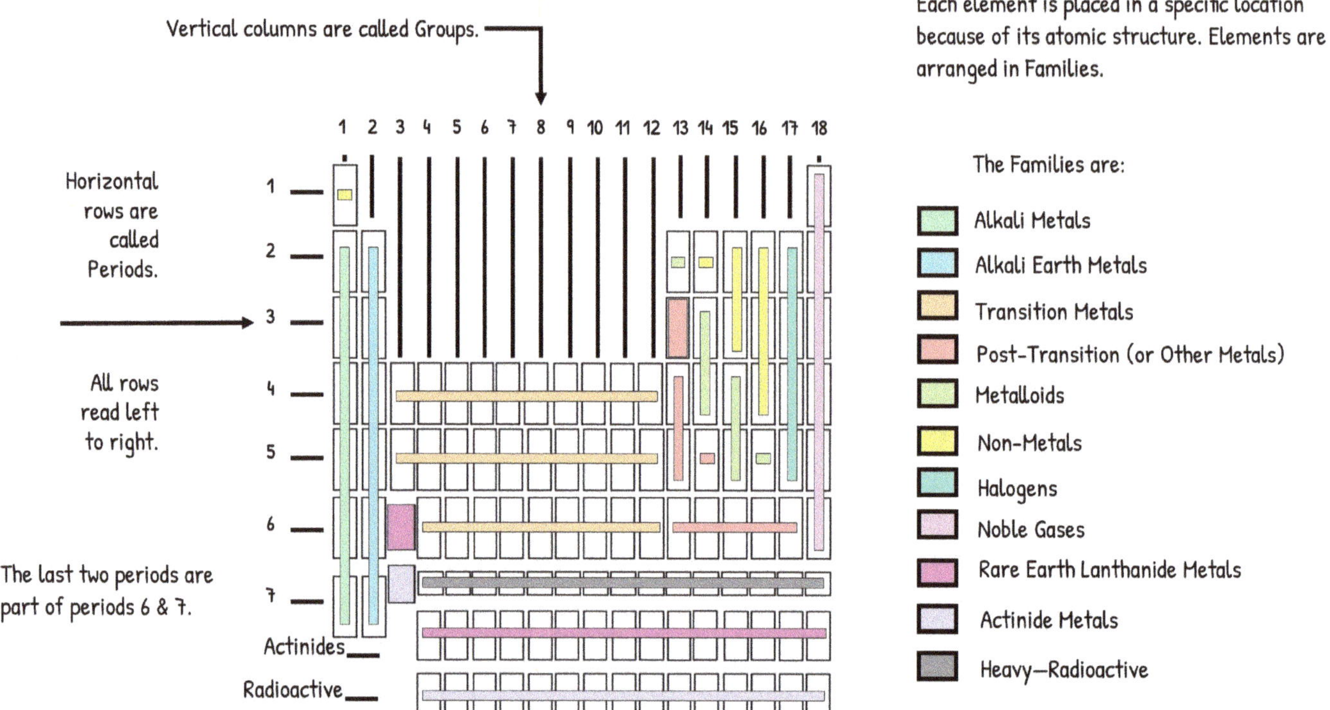

The term 'Element' is used to describe atoms with specific characteristics.
Every element in the first column or Group has 1 electron in the outer orbital (shell).
Every element in the second column (group two) has two electrons in the element's outer orbital.
The number designation of each Group represents the number of electrons in the element's outer orbital—
except for Group 18, Period 1—Helium. It only has 2 electrons.
Those electrons, called Valence Electrons, are what chemically bond with other elements.

Atomic Structure of Element: The atomic structure of an element refers to the arrangement of protons and neutrons in the nucleus of the atom, and the electrons in the electron cloud around the nucleus. Group 1, Period 1—Hydrogen is the only element that has no neutrons.

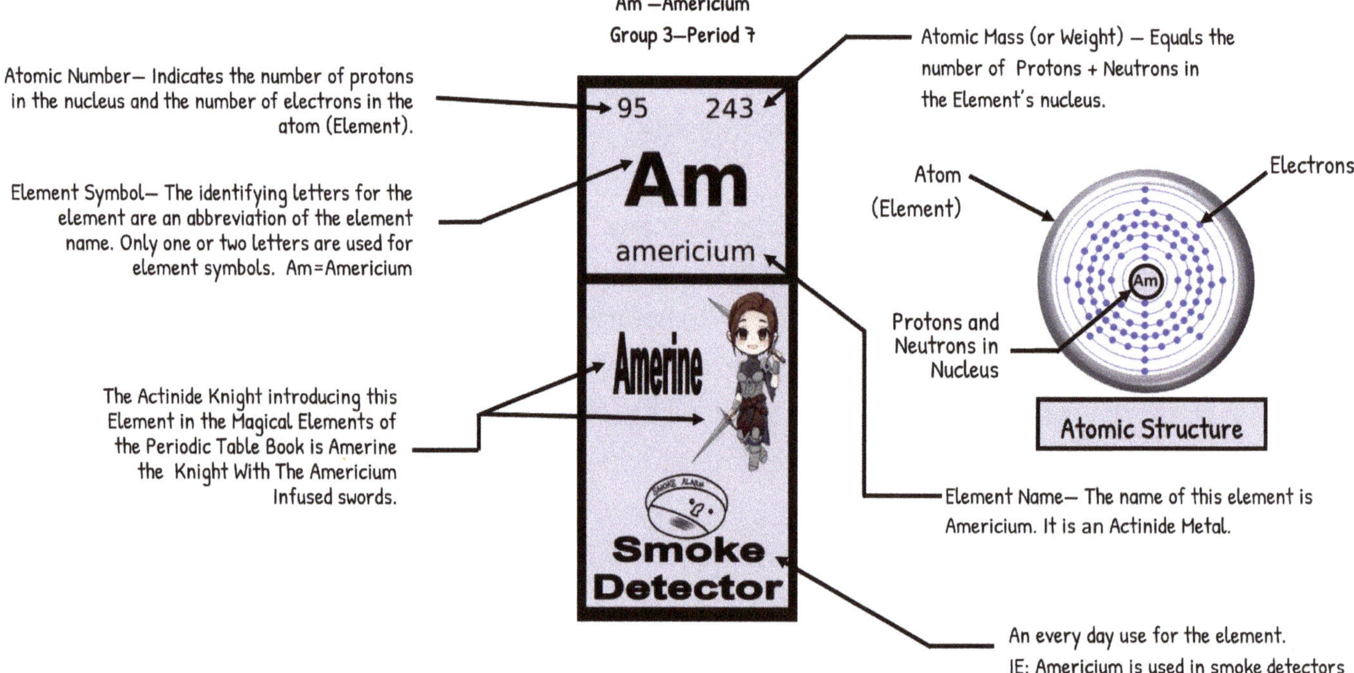

Atomic Number— Indicates the number of protons in the nucleus and the number of electrons in the atom (Element).

Element Symbol— The identifying letters for the element are an abbreviation of the element name. Only one or two letters are used for element symbols. Am=Americium

The Actinide Knight introducing this Element in the Magical Elements of the Periodic Table Book is Amerine the Knight With The Americium Infused swords.

Atomic Mass (or Weight) — Equals the number of Protons + Neutrons in the Element's nucleus.

Element Name— The name of this element is Americium. It is an Actinide Metal.

An every day use for the element. IE: Americium is used in smoke detectors

Types of Elements On The Periodic Table

Alkali Metals—Some metals on the periodic table are soft and shiny. They are so soft that they can be cut with a knife! These metals are excited to give away electrons to elements in need, making them highly reactive. Ther electron transfer creates a compound known as a salt. Surprisingly, these metals are not found in nature alone; they must be extracted from other sources. Examples of these metals include lithium, sodium, potassium, rubidium, cesium, and francium.

Alkali Earth Metals—The elements in column 2 of the periodic table have 2 outer electrons in their shell. Ther makes them very active with nonmetals that need electrons to stay stable. When they react, they make something called a salt. They are often found in nature all by themselves, and they can even conduct electricity. The elements are beryllium, magnesium, calcium, strontium, barium, and radium.

Post-Transition (or other Metals)— Elements directly to the right of the transition metals. They are known as "poor metals: and are soft and brittle. These include aluminum, gallium, indium, tin, thallium, lead, bismuth, zinc, cadmium and mercury.

Transition Metal—The main metals are found in the middle and bottom rows of the periodic table. They look like metal, can conduct electricity, can bend and be shaped easily. The period 4 transition metals are scandium, titanium, vanadium, chromium, manganese, iron, cobalt, nickel, copper, and zinc. The period 5 transition metals are yttrium, zirconium, niobium, molybdenum, technetium, ruthenium, rhodium, palladium, silver, and cadmium. The period 6 transition metals are lanthanum, hafnium, tantalum, tungsten, rhenium, osmium, iridium, platinum, gold, and mercury. The period 7 transition metals are the naturally-occurring actinium, and the artificially produced elements rutherfordium, dubnium, seaborgium, bohrium, hassium, meitnerium, darmstadtium, and roentgenium.

Metalloids—The elements called metalloids are a mix of metals and nonmetals. They look like metals, but can't conduct electricity very well. They also break easily and act like nonmetals. These include boron, silicon, germanium, arsenic, antimony, tellurium, astatine, and polonium.

Non-Metals—These elements reside in columns 15-17, and can be gases, liquids, or solids. They don't conduct heat or electricity. The solids are brittle, and they have no metallic luster. They readily accept electrons from metals to form salts. These include nitrogen, oxygen, fluorine, chlorine, bromine, and iodine.

Halogens—Halogen chemicals are a special type of element. When they mix with metal, they become a kind of salt. Halogens are super reactive because they like to take an electron from metals. They can be found in column 17 of the element table. Some of them can be found in nature, but most are very dangerous and can hurt you if you touch them. They include fluorine, chlorine, bromine, iodine, and the radioactive elements astatine and tennessine.

Noble Gases—These elements reside in column 8. They are all odorless, colorless gases that are chemically very stable (inert). They don't generally form compounds by bonding with another element. These include helium, neon, argon, krypton, xenon, and radon.

Lanthanide Rare Earth Minerals—The Japanese call them "the seeds of technology." The US Department of Energy calls them "technology metals." These elements have atomic numbers 57-71. They are vital to industry. They can be added to metals to strengthen them to make alloys such as stainless steel, used to refine crude oil, and are crucial in producing technology—electronics, telecommunications, and metal devices to name a few. They are lanthanum, cerium, praseodymium, neodymium, promethium, samarium, europium, gadolinium, terbium, dysprosium, holmium, erbium, thulium,

Actinide Metals—Any of a series of chemically similar metallic elements with atomic numbers ranging from 89 (actinium) to 103 (lawrencium). All of these elements are radioactive, and two of the elements, uranium and plutonium, are used to generate nuclear energy. The lanthanides and actinides are sometimes called the inner transition metals, referring to their properties and position on the table. They are actinium, thorium, protactinium, uranium, neptunium, plutonium,

Super Heavy—Radioactive—Superheavy elements are those elements with a large number of protons in their nucleus. Elements with more than 92 protons are unstable; they decay to lighter nuclei with a characteristic half-life. They do not occur in large quantities (if at all) naturally on earth, and only exist briefly under highly controlled circumstances. They include lawrencium, rutherfordium, dubnium, seaborgium, bohrium, hassium, meitnerium, darmstadtium, roentgenium, copernicium, nihonium, flerovium, moscovium, livermorium, tennessine, and oganesson.

Alpha-Emitting Transuranic Actinide Elements

These elements are all synthetic, meaning they don't occur naturally on Earth, and are produced through nuclear reactions. They are all radioactive and decay by emitting alpha particles, among other decay modes, Transuranic or transuranium elements can be classified as technogenic nuclides, meaning it has been produced and released into the environment due to human nuclear activity.

These four elements are the most abundant and the most extensively used of the man-made actinide series elements. They comprise a major radioactive waste disposal concern because they are long lived and have high radiotoxicity. Neptunium and Plutonium are the only transuranium elements that have been found in trace amounts in nature.

Neptunium
2,100,000 Year Half Life

Plutonium
24,000 Year Half Life

Americium
458 Year Half Life

Curium
17.6 Year Half Life

These seven elements are produced in such small amounts, mostly for research purposes; and most of the isotopes produced have such short half-lives, a few seconds or minutes, that they are an unlikely health concern.

Berkelium   Californium   Einsteinium   Fermium

  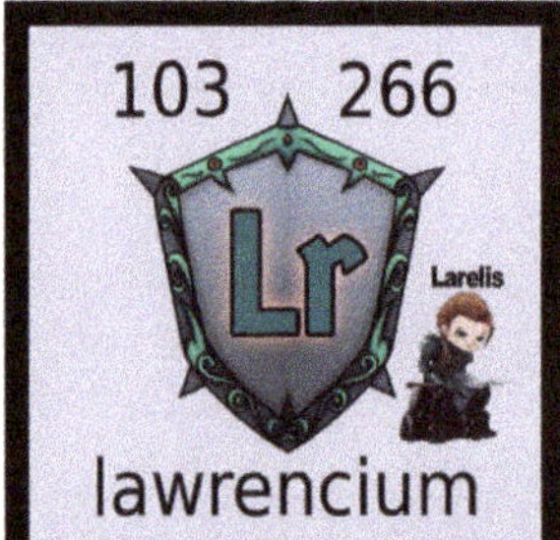

Mendelevium   Nobelium   Lawrencium

Radioactive Isotopes

All elements have **isotopes**. There are Stable Isotopes and there are Radioactive Isotopes which are known to be Unstable. Every chemical element has one or more radioactive isotopes. More than 3,000 Radioactive Isotopes (radioisotopes) are known, of which only about 84 are seen in nature. We will discuss some of them here.

Uranium =

3 Naturally Occurring Radioisotopes:

Uranium-238

Uranium-235

Uranium-234

Uranium has 24 man-made radioisotopes.

<u>U-238</u> Radioisotope Applications: The half-life of uranium-238 is approximately 4.468 billion years. It's alpha decay begins with thorium-234. This is the most common isotope of uranium found in nature. It can be used to generate plutonium-239, which itself can be used in a nuclear weapon or as a *nuclear-reactor* fuel supply.

Radium =

4 Naturally Occurring Radioisotopes:

Radium-223

Radium-224

Radium-226

Radium-228

Radium has 34 man-made radioisotopes.

<u>Radium-223</u> Radioisotope Applications: Has a half-life of 11.43 days and is part of the actinium decay series. It *targets bone tissue with alpha particles*, which have a short path length, reducing toxicity to adjacent healthy tissue. It can control painful bone metastases, delay complications like fractures, and improve survival.

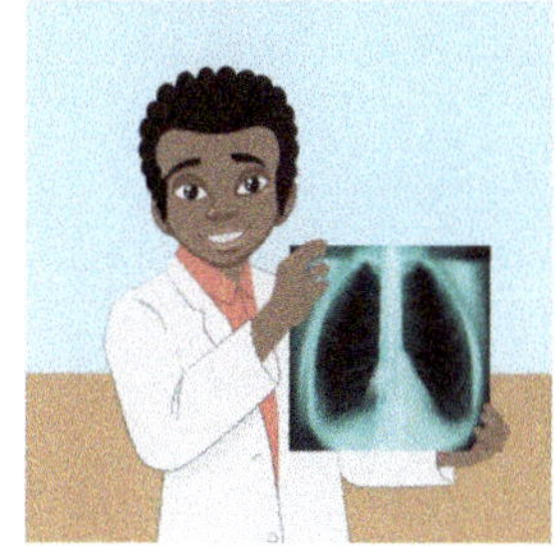

Francium =

2 Naturally Occurring Radioisotopes:

Francium-221

Francium-223

Francium has 32 man-made radioisotopes

<u>Francium-221</u> Radioisotope Applications: Francium-221 has a half-life of 4.8 minutes. When it undergoes alpha decay, it loses 2 protons and 2 neutrons, resulting in the formation of astatine-217. Due to its rarity and extreme instability, has no commercial applications and is primarily used for research purposes in fields like chemistry and atomic structure, including *spectroscopic investigations*.

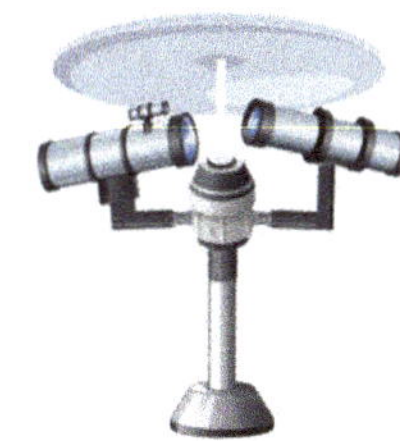

Polonium =

4 Naturally Occurring Radioisotope:

Polonium-210

Polonium has 41 man-made radioisotopes

<u>Polonium-210</u> Radioisotope Applications: Polonium-209 has the longest half-life of all polonium isotopes at 124 years. It is formed as a decay product of uranium-238. It is primarily used in industrial devices to eliminate static electricity, particularly in processes like *paper rolling*, plastic sheet manufacturing, and spinning synthetic fibers. It's also used in neutron sources and dust removal brushes for photographic films and camera lenses

Can you guess the most commonly used radioisotope?

= ??????

(Answer can be found below.)

The above chart only shows a few of the radioactive isotopes formed by those elements. Other very important radioisotopes are:

Iodine-131: Used to diagnose and treat various diseases associated with the human thyroid. **Molybdenum-99 (Mo-99):** Used as the 'parent' in a generator to produce technetium-99m. **Americium-241 (Am-241):** Used in smoke detectors, and other applications like thickness gauges and neutron sources.

The most commonly used radioisotope is *Technetium-99m (99mTc)*, employed in over half of all nuclear medicine procedures, used for imaging various organs and tissues. It is used in tens of millions of medical diagnostic procedures annually

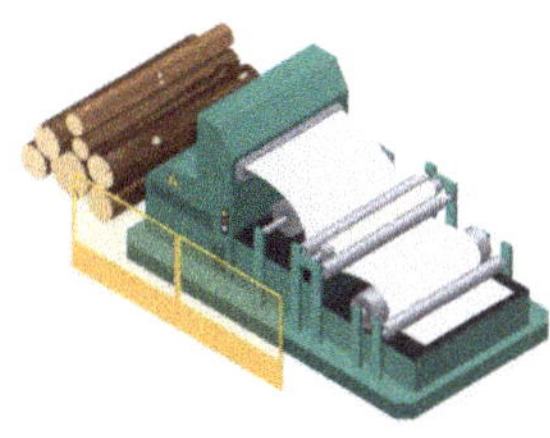

Definitions

Atomic Structure of Element: The atomic structure of an element refers to the arrangement of protons and neutrons in the nucleus of the atom, and the electrons in the electron cloud around the nucleus.

Atomic Number: An element's atomic number refers to the number of protons it has in its nucleus. In a neutral atom the number of protons always equals the number of electrons.

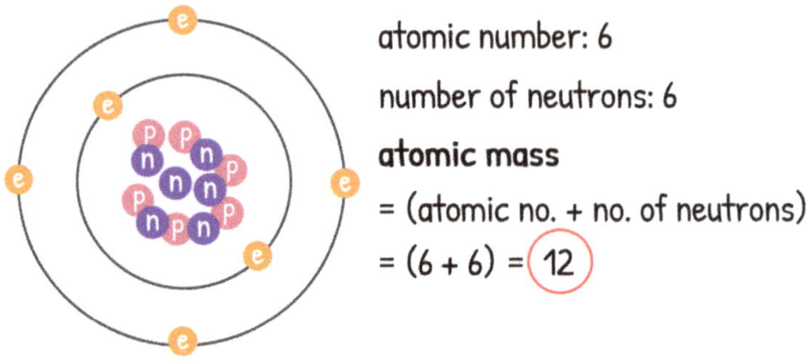

Carbon atom

Atomic Weight (Mass) of Element: The atomic mass of an element is how heavy it is. It is made up of protons and neutrons that are in the middle of the element. Some elements have different versions with different amounts of neutrons, but they still have the same amount of protons. The atomic mass is the average weight of all these versions of the element.

Allotrope: Allotropes are different forms of an element that look and act different, but are made of the same stuff. Some elements have more than one form. For instance, carbon can be a shiny diamond or a gray pencil lead called graphite.

Isotope: Isotopes are different types of atoms that have the same parts, like protons and electrons, but they have a different number of neutrons. For example, the three most stable isotopes of hydrogen: protium (A = 1), deuterium (A = 2), and tritium (A = 3).

Crystalline Structure of Element: The crystalline structure of an element is how its atoms, ions, or molecules stick together in a pattern to make a cool crystal shape.

Ferrous and Non-Ferrous Metals: When we say ferrous metal, it means that iron is a big part of the metal. But if there's only a little bit of iron in the metal, we call it non-ferrous. The word "ferrous" comes from Latin and means iron, which is why iron's symbol is Fe.

Ductile Metals: These are capable of being made into long, thin wire or thread. Copper and Silver are ductile metals.

Malleable Metals: These can be hammered or rolled into thin sheets without cracking or breaking. Gold is malleable.

Ferromagnetic: Materials that are strongly attracted to a magnet. Such materials can be permanently magnetized. These include the elements iron, nickel and cobalt and their alloys, some alloys of rare-earth metals, and some naturally occurring minerals such as lodestone.

Magnetostriction—Ther is the term for a special thing that happens to magnetic materials. When these materials get turned into magnets, they also change their shape or size.

Paramagnetic: Slightly attracted to a magnetic field, but do not retain magnetic properties once the field is removed.

Diamagnetic: Slightly repelled by a magnetic field, but do not retain magnetic properties once the field is removed.

Electrical Properties: Conductor—a thing that lets electricity flow through it. Semi-conductor—a special material (usually silicon) that can conduct electricity, but not as well as metal. Insulator (non-conductor)—a material (usually glass) that stops electricity from flowing.

Reactive Gas: These gases are really good at reacting with stuff! They are called "sticky gases" because they can react to things like plastic and wet surfaces when they touch them. These are nitrogen, oxygen, hydrogen, carbon dioxide, fluorine, and chlorine.

Non-Reactive Gas: An inert gas is like a super shy gas that doesn't like to hang out with other chemicals. It doesn't make any new friends by reacting with them, so it doesn't form any chemical compounds. We also call these special gases "noble gases."

What Makes Americium Seem Magical?

Americium would probably seem amazing, strange, and even a little scary to people from long ago. It does things that sound like magic, even though we know they are real. One of the most unusual things about it is that it can glow in the dark. Not like a lantern or a fire, but in a quiet, ghostlike way that feels eerie. If an ancient person saw that, they might think the stone or metal was cursed, blessed, or alive.

It is also used in smoke detectors, where it helps find smoke very quickly. It works without making noise and without being seen, like a hidden guard that never falls asleep. To people who lived before electricity and science, that would be hard to believe. They might say it was some kind of spirit watching over a home, warning people before danger arrived.

Another reason americium would have seemed magical is that people make it. It does not just come out of the ground in large amounts. Humans create it in nuclear reactors by changing one kind of atom into another. Long ago, this would have sounded like alchemy, the old dream of turning one substance into something else. For many people in ancient times, that alone would have seemed impossible.

Americium is also powerful in a way that cannot be seen. It gives off radiation that you cannot smell, hear, or touch, but it can still affect the world around it. That kind of hidden force would have been very hard for early people to understand. They might have thought it was an invisible flame or a secret energy from the stars.

It is useful in ways that could look like tricks from a wizard's workshop. Smoke detectors, measuring tools, and other devices use americium to do careful jobs that help keep people safe. A small amount of it can do something very important, which would have seemed like a miracle to people from the past.

Americium also belongs to the age of atoms, because it is an element beyond uranium, one of the heavier elements known to science. The idea that people found and made something so deep and hidden in nature would have felt like uncovering a secret part of the world.

So to our ancestors, americium might have seemed like a forged moonstone with invisible fire inside it: a man-made material with strange power, a soft ghostly glow, and the power to warn of danger before humans even know it is there.

Potential Future Uses For Americium

Americium is a strange and interesting metal. Most people know it only because it is used in some smoke detectors, but in the future it may have other jobs too. It was first made in a lab, so it is not found in large amounts in nature. That makes it feel a little mysterious. Still, scientists often find new uses for materials that once seemed useful only in one small way.

One possible future use for americium is in tiny power sources. Americium gives off heat as it slowly changes over time. That heat could one day be used to power very small machines that need to work for a long time without new batteries. These machines might be placed in far-away places, like deep caves, under the sea, or on other planets. A small device with americium might keep working for years, even where sunlight and regular batteries would not help much.

Another possible use is in space travel. Space ships and space tools need power when they are far from the Sun. If scientists can make americium safer and easier to use, it could help power small probes or sensors. It could also support tools that study planets, moons, or asteroids. In space, weight matters a lot, so a tiny source of steady energy could be very valuable.

Americium may also help in science and medicine. Its special radiation could be used in careful ways to study materials, find hidden cracks, or test how strong something is. In the future, it might help make better machines that look inside objects without breaking them open. Doctors may also use related ideas in a very careful and limited way to find disease earlier or guide treatment, if safer methods are found.

There is also a chance that americium could help with long-lasting sensors. Imagine a sensor buried in a bridge, a volcano, or a frozen place in the far north. It could send back information for many years without needing a battery change. That would be useful for safety, weather study, and building research. A tiny americium source could make such devices last much longer than today's regular batteries.

Of course, americium is not easy to use. It must be handled with great care because it gives off radiation. Future scientists will need to make sure it is safe, controlled, and never wasted. But if they solve these problems, americium could become a quiet helper in space, science, and remote technology. Something once seen as a small, odd metal could end up doing very big work.

Meet Amerine, The Knight With The Americium Tipped Swords

Amerine had grown up in MarBryn, where the valleys were green, the winters were long, and people learned early that safety was something worth building and defending. She studied science, metalwork, and the old village reactor systems that kept Kilmere warm and bright. When she was older, she became known as the Actinide Knight, a protector who carried twin Americium blades—weapons of controlled power, made to defend without causing ruin.

Glaros had once been a learned man too, but anger changed him. He believed power should be forced, not shared. He wandered from place to place, leaving fear behind him, until he came to Kilmere and saw the reactor as a prize. If he could seize its energy, he could turn the whole valley into a weapon. That was why Amerine was already on the move when the first smoke rose.

She crossed the edge of the valley with a quick, steady pace, her armor catching the early light. Below her, Kilmere shone softly, its walls golden and its cottages close together beneath cobalt banners. The air smelled of pine and thawing water, but under that calm lay a warning. Amerine knew how to feel for danger before it fully arrived.

By the time she reached the village square, fear had already spread. Lanterns flickered with a strange green glow, and the people stood tense and silent. Then Glaros appeared on the eastern ridge, tall in ash-gray robes, his staff planted in the ground like he owned the place. Dark energy curled around his hands as he began a chant that made the air feel heavy.

Amerine stepped forward with her twin blades lowered, so the villagers would know she was there to protect them. She saw the old grandmother clutching her shawl, the boy with a worn slingshot, and the craftsman with soot on his sleeves. They were frightened, but they were not broken. She would not let Glaros take that from them.

Glaros raised his staff, and a ring of blue fire burst from the tip, cold and sharp instead of warm. A cruel smile touched his mouth. "Do you see what I've done, Knight?" he rasped. "The reactor below this village can be bent to my will."

Amerine's right hand tightened on her sword. The blade hummed with pale light. The second blade answered with a softer tone, and the two sounds blended in a steady, ready harmony. "This ends now, Glaros," she said.

He laughed. "Only the weak speak of endings. I will shackle this village with fear."

Blue rings shot toward her, meant to bind her arms and legs. Amerine leapt aside with the grace of a dancer, her blades flashing as they cut the magic apart. The chains broke into harmless sparks that drifted to the ground like snow.

Glaros sent another wave of force at her, trying to push her back toward the reactor access below the square. Amerine kept moving, but not wildly. She had learned restraint long ago. Her blades were powerful, but they were meant to cut danger away, not spread it. She used careful footwork and precise strikes to keep the villagers safe while driving him toward the center of the square.

The sorcerer threw up phantom shapes—false shadows meant to frighten the crowd. Amerine saw through them. She knew his power depended on fear, and she would not give him any. With every bright slash, the false images broke apart, and the air seemed clearer.

Beneath the village, the reactor hummed steadily, protected by safeguards Amerine had helped install during her service. She knew every safety system, every barrier, every line of energy that held Kilmere together. That knowledge gave her strength now. When Glaros lunged with a dark surge meant to drain the reactor dry, she dropped into a deep lunge and turned his spell away from the village.

The vortex swept out across the open fields instead, where it lost strength and died harmlessly in the earth. Glaros staggered and fell to one knee, his power spent.

Amerine did not strike him down. Instead, she placed one glowing blade at his throat and the other against his shoulder. "Leave Kilmere," she said. "Leave the reactor. If you return, I will be waiting."

For a moment, Glaros said nothing. Then his hard expression broke, and he looked not like a conqueror, but like a man who had lost his way. "What is power if it cannot protect?" he whispered.

Amerine let him go. He stumbled back into the shadows, and the villagers rushed toward her with tears, relief, and praise. The grandmother touched her gauntlet and offered a blessing. The boy stood a little taller, as if he had found courage of his own.

By morning, Kilmere was quiet again. The reactor's hum had settled into a steady, reassuring rhythm, and the village began to repair what fear had damaged. Amerine helped where she could—fixing safety systems, teaching blade care, and reminding people that strength meant little without mercy.

As evening came, she stood at the edge of the village and looked toward the ridge where Glaros had disappeared. She knew danger would return someday. But she also knew Kilmere was strong, and that courage, discipline, and kindness could hold the valley together.

Amerine turned back toward the village with a faint smile. Her armor caught the last light of day, and her blades glowed softly at her sides. She was not only a warrior. She was a guardian. And in Kilmere, where the lights now shone a little brighter, the people knew that as long as the Actinide Knight stood watch, they would not face the dark alone.

Enjoy This Coloring Page Featuring

Amerine The Knight With The Americium Tipped Swords

Sample Page From Magical Elements of the Periodic Table

Presented By The Actinide Knights

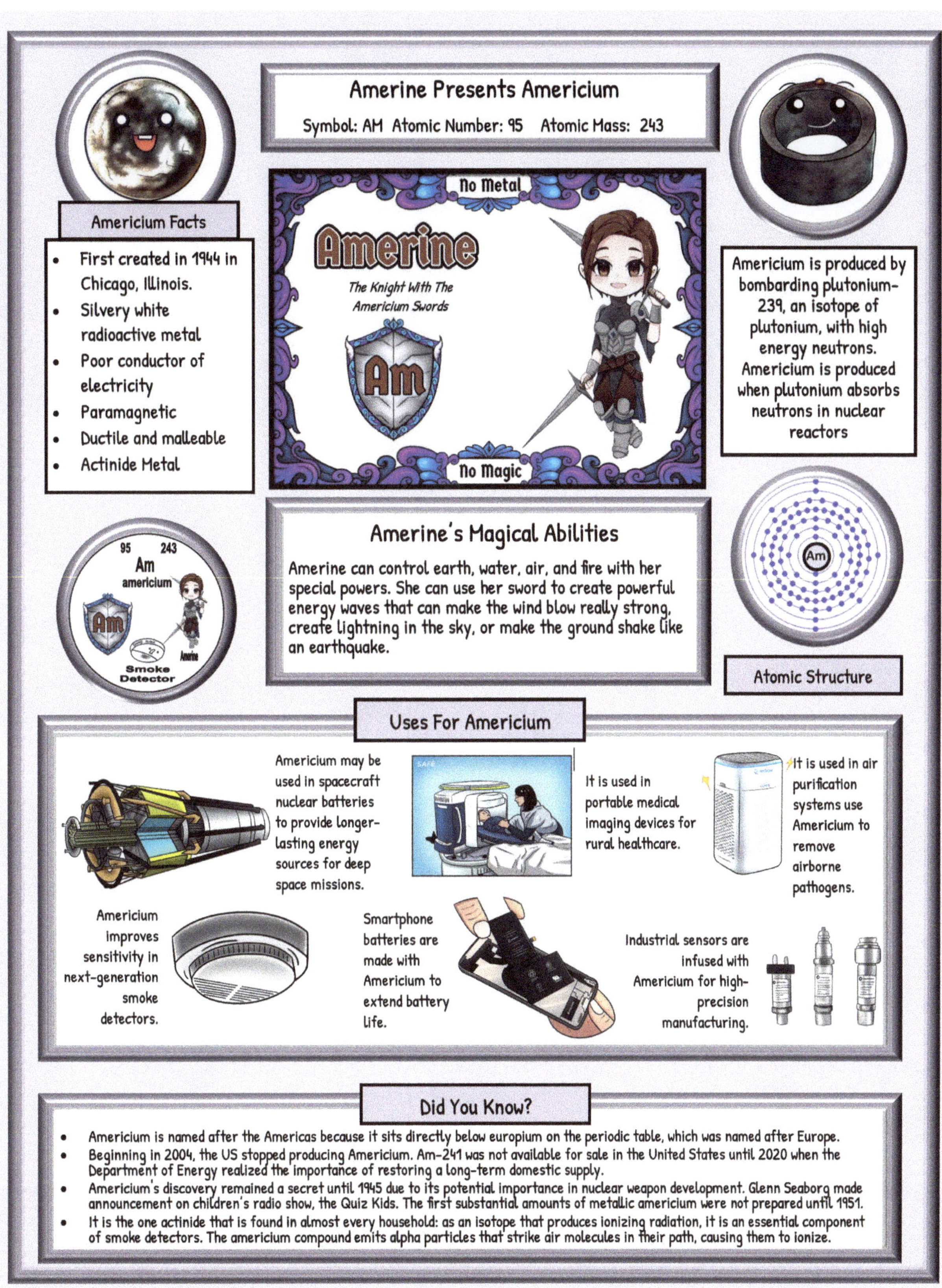

MEET THE ACTINIDE KNIGHTS

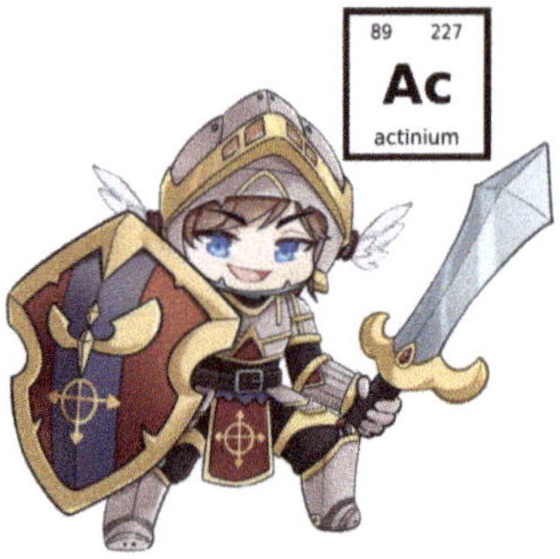

Acamus—Actinium
(Not a Transuranic Element)

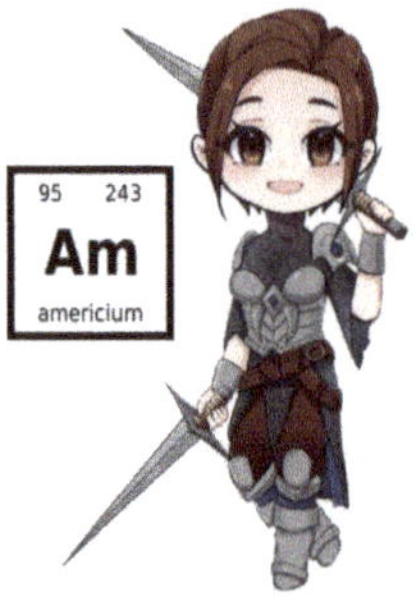

Amerine—Americium
(Transuranic Element)

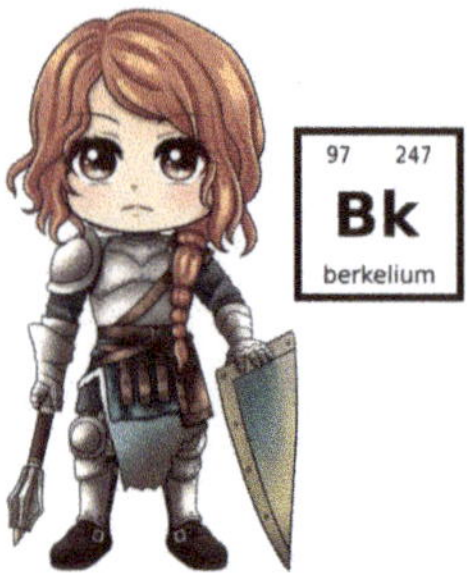

Beremere—Berkelium
(Transuranic Element)

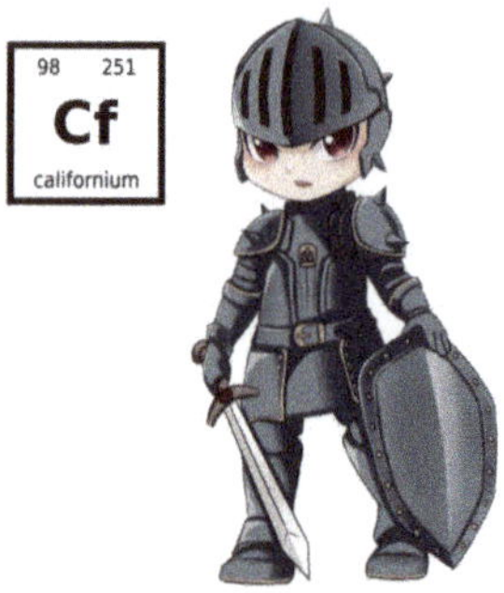

Calastian—Californium
(Transuranic Element)

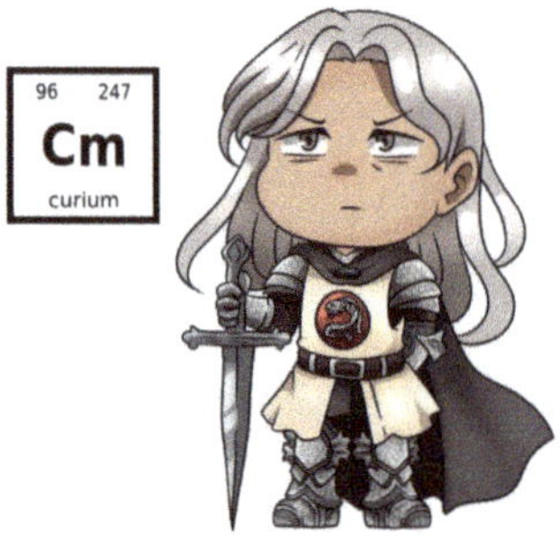

Curran—Curium
(Transuranic Element)

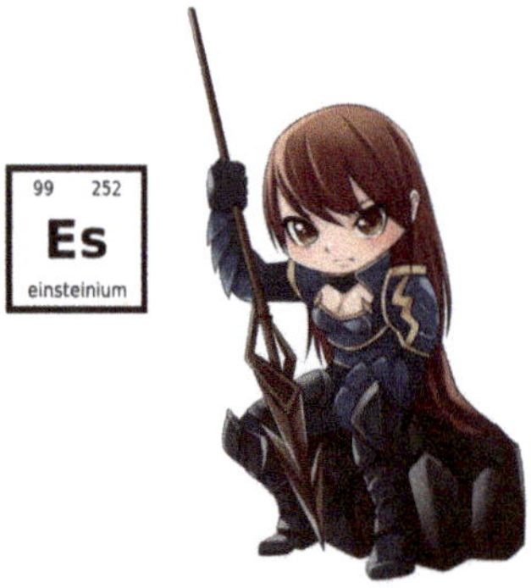

Elizama—Einsteinium
(Transuranic Element)

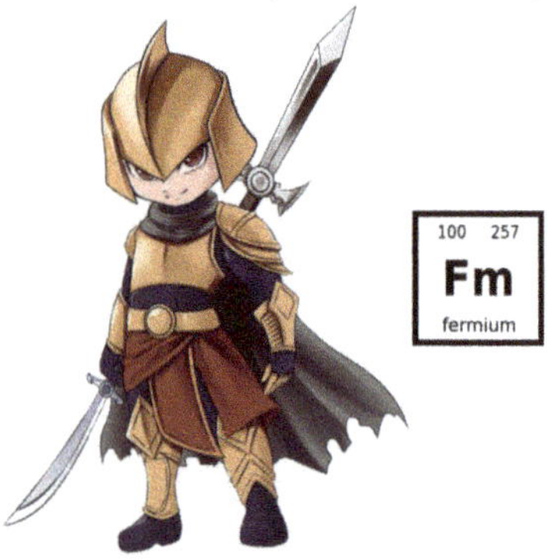

Ferley—Fermium
(Transuranic Element)

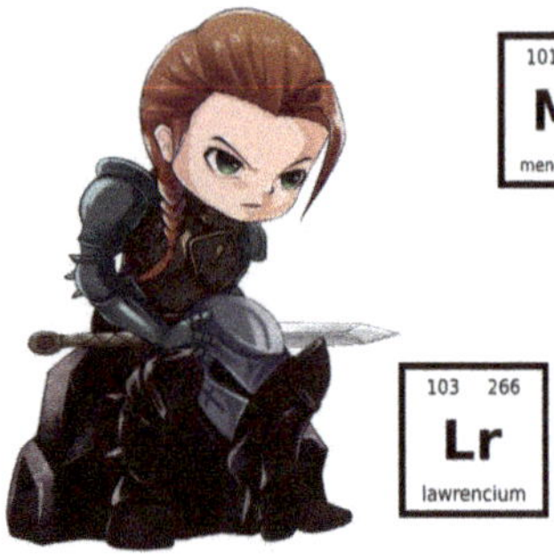

Larelis—Lawrencium
(Transuranic Element)

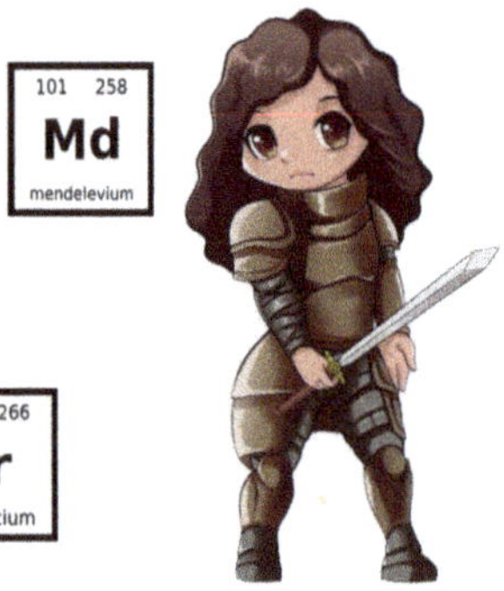

Menesant—Mendelevium
(Transuranic Element)

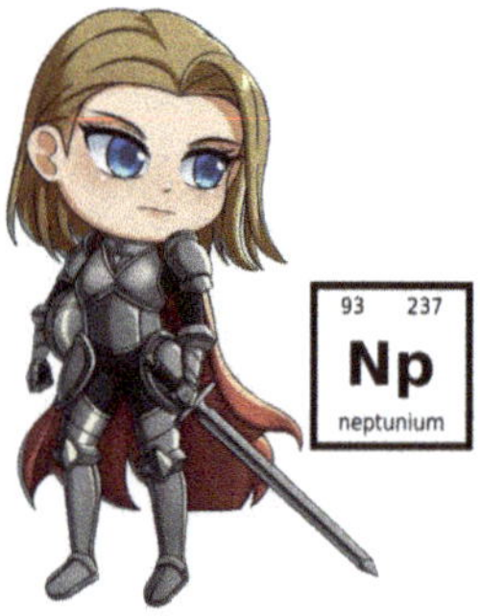

Nepthas—Neptunium
(Transuranic Element)

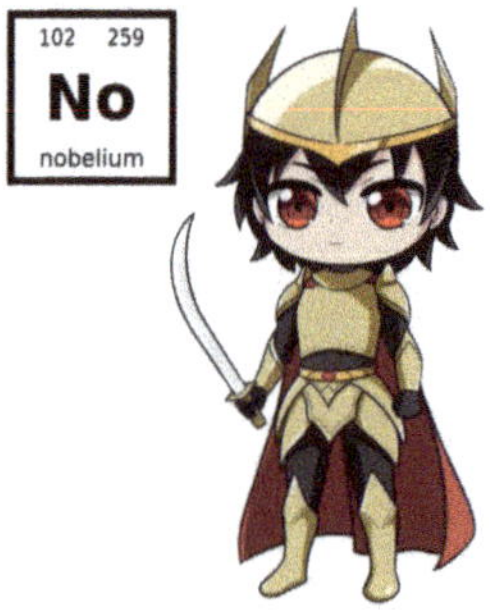

Tellan-Tellurium
(Transuranic Element)

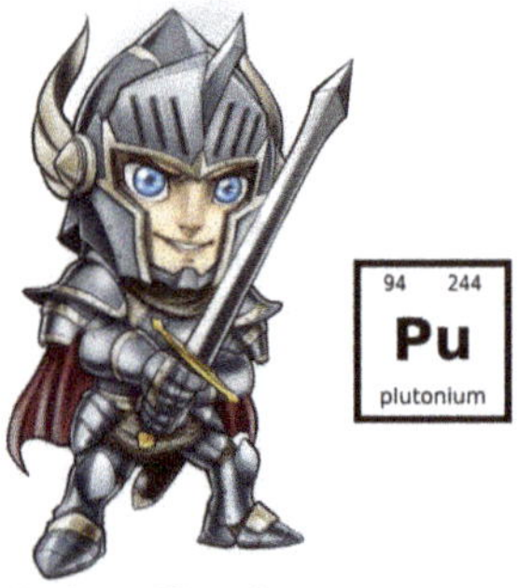

Puchan—Plutonium
(Transuranic Element)

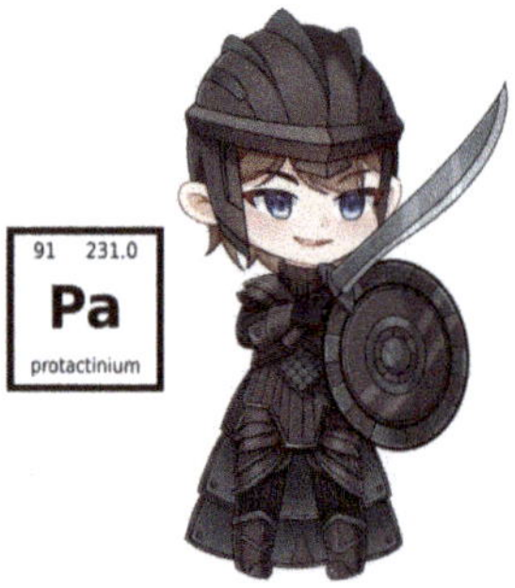

Protik—Proactinium
(Not a Transuranic Element)

Thordis—Thorium
(Not a Transuranic Element)

THEY WIELD SWORDS, LANCES AND MACES POWERED BY ELEMENTS FROM THE PERIODIC TABLE

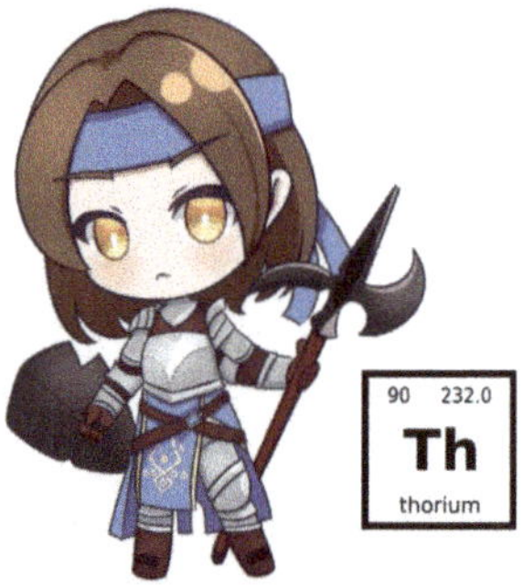

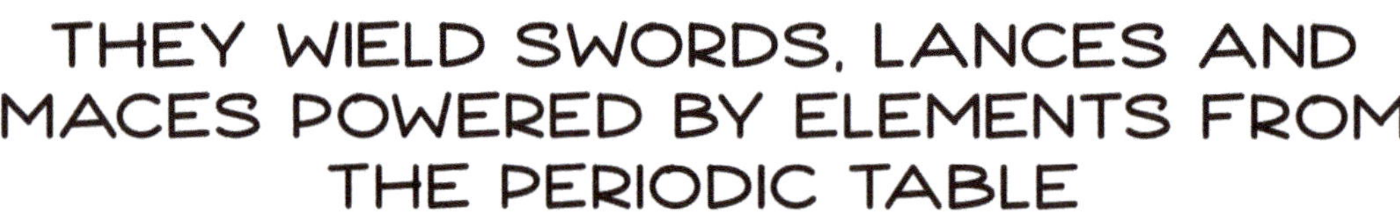

Magical Elements of The Periodic Table

Create Your Own Magical Actinide Knight Elemental

Amerine — The Knight With The Americium Swords

Symbol: Am Atomic Number: 95 Atomic Mass: 243

Magical Elemental Symbol

Produced by bombarding plutonium with neutrons

Atomic Structure

95 243
Am
americium

Amerine's Magical Abilities

She can use her sword to create powerful energy waves that can make the wind blow really strong, create lightning in the sky, or make the ground shake like an earthquake.

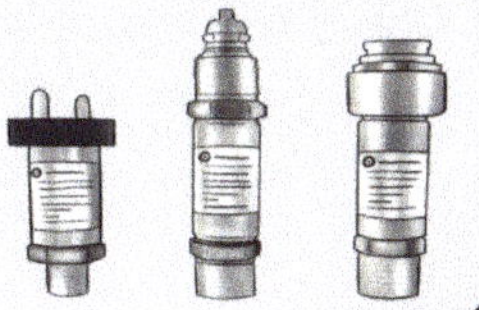

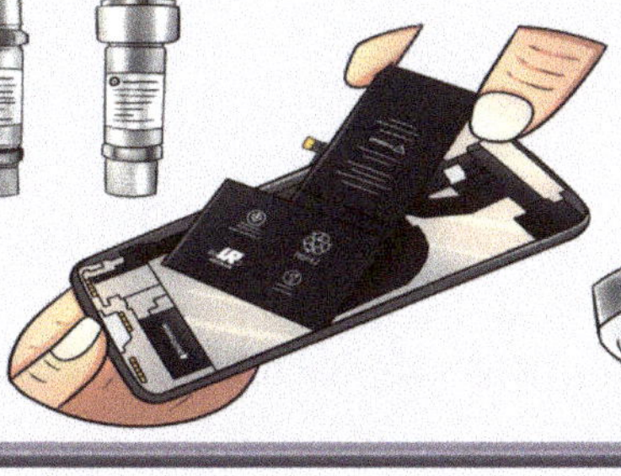

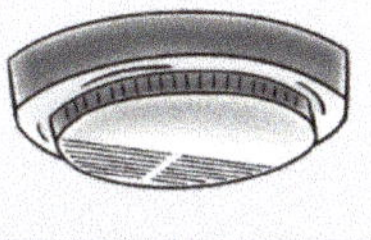

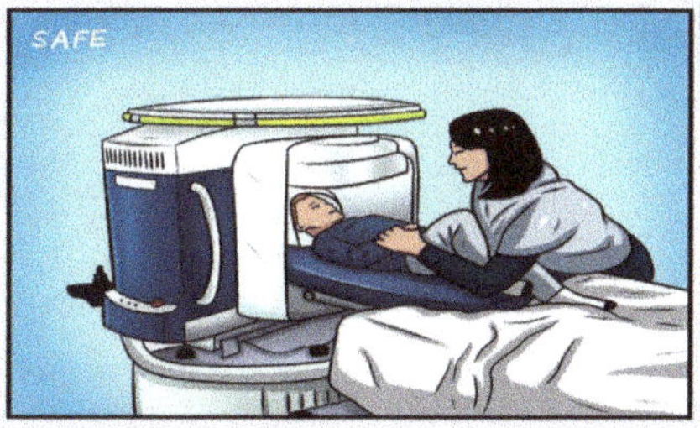

Americium is a Actinide Metal

95 243
Am
americium
Am
Smoke Detector

Americium Periodic Symbol

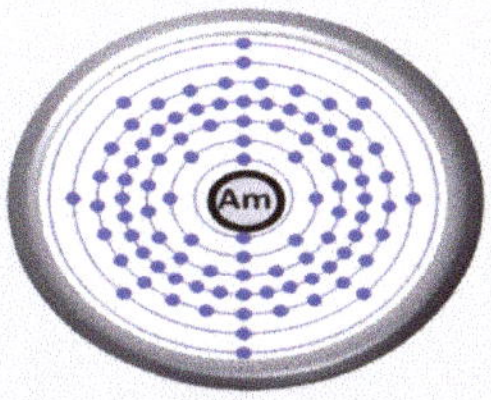

Magical Elements of The Periodic Table

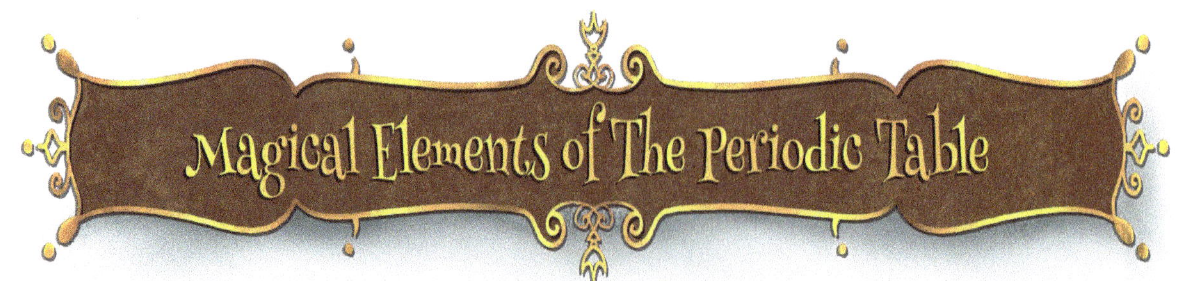

Students may either use a program like power point to cut and paste clip art into a Magical Knight Elemental Blank or, if they wish, they may draw everything themselves.

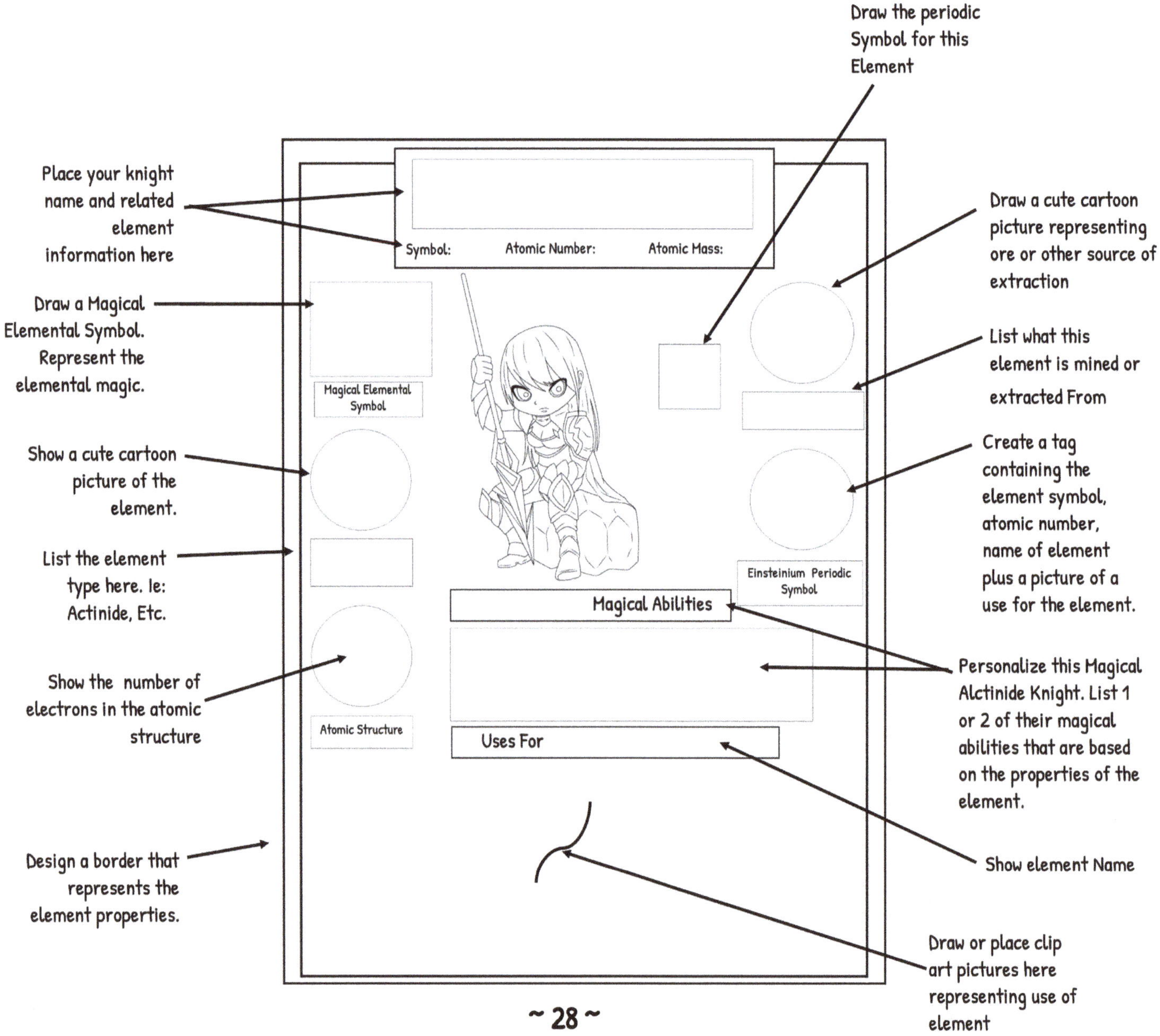

Draw the periodic Symbol for this Element
Place your knight name and related element information here
Symbol:
Atomic Number:
Atomic Mass:
Draw a cute cartoon picture representing ore or other source of extraction
Draw a Magical Elemental Symbol. Represent the elemental magic.
Magical Elemental Symbol
List what this element is mined or extracted From
Show a cute cartoon picture of the element.
Create a tag containing the element symbol, atomic number, name of element plus a picture of a use for the element.
List the element type here. Ie: Actinide, Etc.
Einsteinium Periodic Symbol
Magical Abilities
Show the number of electrons in the atomic structure
Atomic Structure
Personalize this Magical Alctinide Knight. List 1 or 2 of their magical abilities that are based on the properties of the element.
Uses For
Design a border that represents the element properties.
Show element Name
Draw or place clip art pictures here representing use of element

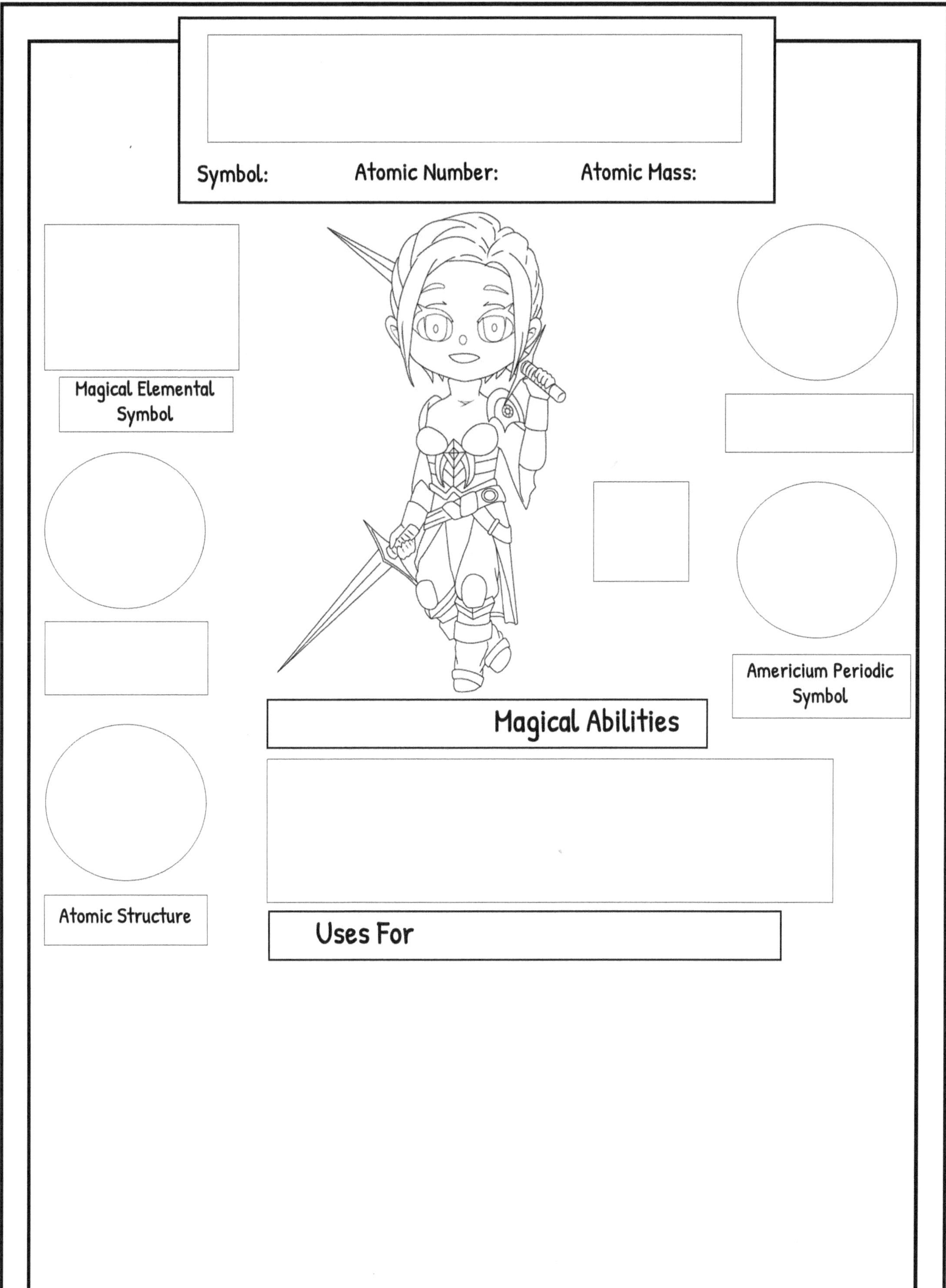

Symbol:
Atomic Number:
Atomic Mass:
Magical Elemental Symbol
Atomic Structure
Magical Abilities
Uses For
Americium Periodic Symbol

Magical Knight Elemental Research Sheet

Before starting your Magical Knight Elemental graphics page, do some research on your chosen element.

Name of Magical Knight:	
Knight's Magic Power Based on the Element's Properties:	
Magical Elemental Symbol:	
Element Name:	
Element Symbol:	
Atomic Number:	
Atomic Mass:	
What year and where was this Element discovered?	
Who discovered this Element?	
Element Group:	
Element Period:	
Element Family Name:	
State of Element At Room Temperature:	
What is Element Mined or Extracted From?	
Is Element Magnetic?	
Does Element Conduct Electricity?	
Where is the Element commonly found in Nature?	
What is 1 alloy of the Element? How used?	
What is 1 compound of the Element? How used?	
Name the most common use for this Element:	
Name a little known use for this Element:	
Name one more use for this Element:	
Interesting and Fun Facts:	

Magical Unicorn Elemental Research Sheet

Before starting your Magical Unicorn Elemental graphics page, do some research on your chosen element.

Name of Magical Unicorn:	Ghel The Gold Horn Unicorn
Unicorn's Magic Power Based on the Element's Properties:	Ghel can see past, present and future. She is empathic and can sympathize with the feelings of other. They say she has a heart of gold.
Magical Herd Crest Symbol:	An open heart with a Celtic Trinity Knot.
Element Name:	Gold
Element Symbol:	Au— Comes from Aurum which is the Latin word for Gold.
Atomic Number:	79
Atomic Mass:	196.97
What year and where was this Element discovered?	Around 4,600 BCE in Bulgaria
Who discovered this Element?	Unknown
Element Group:	11 on Periodic TAble
Element Period:	6 on Periodic TAble
Element Family Name:	Gold is a Noble Transition Metal
State of Element At Room Temperature:	Solid
What is Element Mined or Extracted From?	Quartz Veins. It is also found in gravel in streams.
Is Element Magnetic?	It is Diamagnetic. It's only weakly magnetized when placed in a magnetic field.
Does Element Conduct Electricity?	Gold is a great electrical conductor used in printed circuitry of computers.
Where is the Element commonly found in Nature?	One of the largest deposits is found in the United States in Arkansas.
What is 1 alloy of the Element? How used?	White gold is an alloy of gold, palladium, nickel and zinc.
What is 1 compound of the Element? How used?	Gold Phosphide is a semiconductor used in high power, high frequency applications and in laser diodes.
Name the most common use for this Element:	Jewelry
Name a little known use for this Element:	Acupuncture needles
Name one more use for this Element:	Gold is used in airbags in cars.
Interesting and Fun Facts:	Gold was used in ancient Egypt to fill decayed teeth. Gold thread is incorporated in astronaut spacesuits to protect them from the heat of the sun.

Write a paragraph below to describe your magical knight elemental. Based on the information obtained from research of your chosen element, how did you determine your knight's name? What are your knight's magic powers? What are their likes/dislikes, strengths/weaknesses, personality traits? What color are your knight's armor and clothing and why did you pick that/those colors? What is your knight's Magical Elemental Symbol?

Magical Unicorn Elemental Sample Description

Ghel The Gold-Horned Unicorn

This magical unicorn has a golden horn and hooves that glow like the sun. Her hide is honey-gold and her flowing mane and tail are golden-blonde.

Ghel is a member of the Metal Horn Unicorn Tribe from Unimaise. Gold is linked to the heart chakra because it holds a warm energy that brings soothing vibrations to the body to aid in the healing process. Ghel's Magical Herd Crest symbol is an open heart with a Celtic Trinity knot which indicates that she can see past, present and future. She is empathic and can sympathize with the feelings of others.

Being empathic does not make Ghel a weakling. She is brave with strong opinions and is a true champion to those she loves. She is known as the unicorn with the "heart of gold". When she places her horn on the heart of another, she senses their future.

The name Ghel is an Indo-European word which means yellow. The word "gold" most likely has its origins in the word "Ghel".

Gold is known to possess spiritual powers that bring happiness, peace, stability and luck to those who wear it. Scientists say that all the gold in the world comes from the collision of neutron stars.

Though most other magical unicorn elementals get along well with the gold horn unicorn herd; Ghel, and others like her, must be very careful around the Quick Silver Herd - as gold dissolves in mercury.

Do Your Middle Graders Want To Know More From The Magical Elementals About The Periodic Table?

Get the accompanying books in print at all online book stores. Get the books and accompanying activities at MagicalPTElements.

Available Now

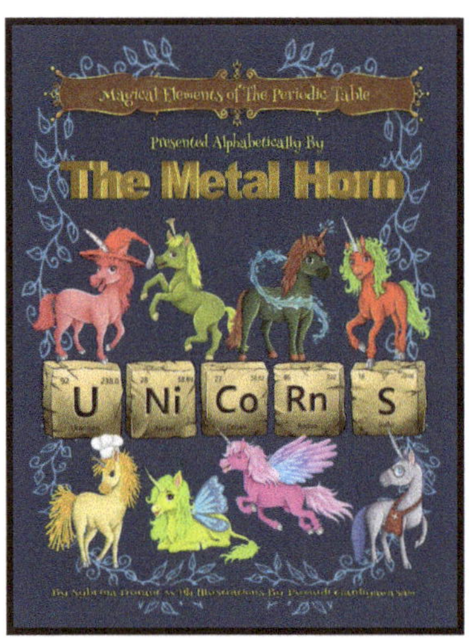

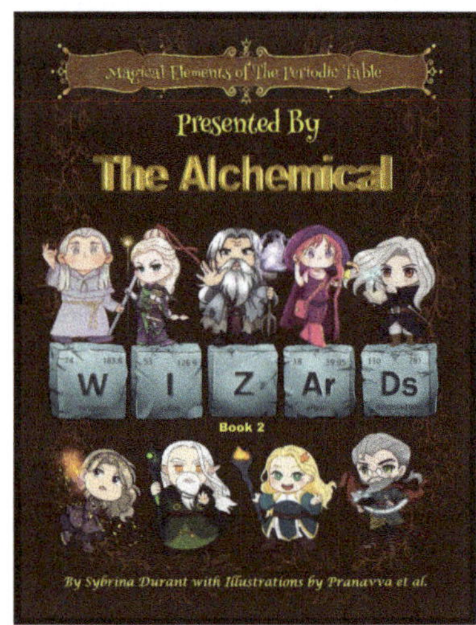

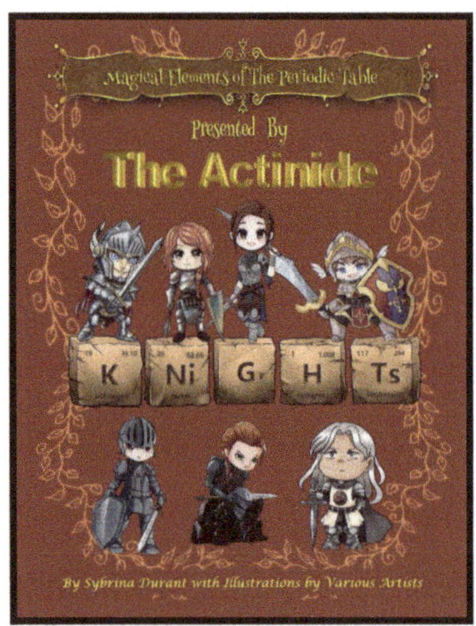

Get These Trading Cards Sets Featuring The Magical Elementals Representing The Periodic Table Elements

at https://bit.ly/40oEUBr

Unicorns, Dragons, Wizards, Knights, or Goblins?

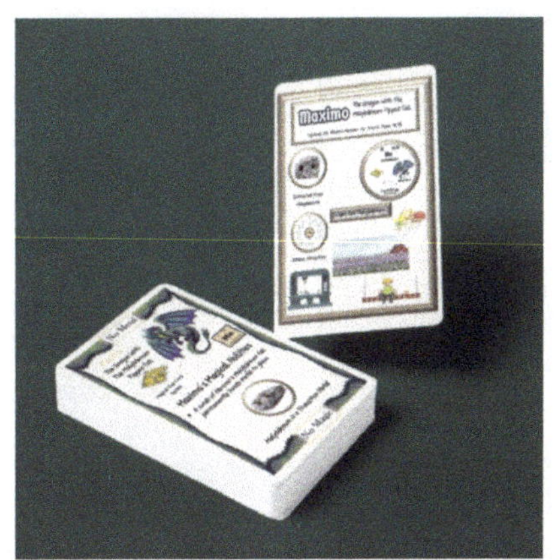

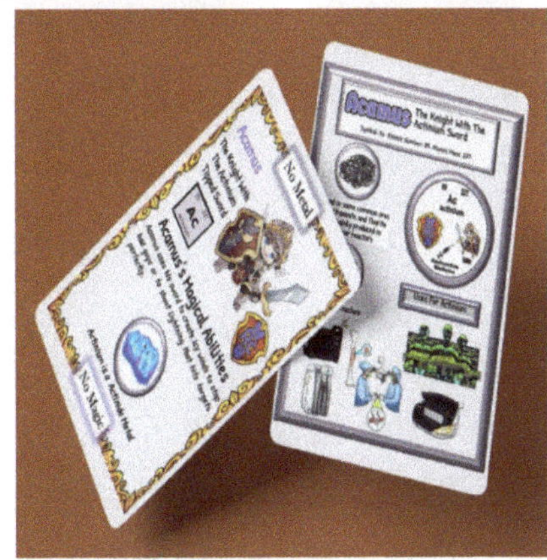

Collect Them All

These magical elementals are ready to help make learning the periodic table more fun. Get all books and related activities today.

Learn More About all of the Periodic Table Elementals.
Get all of the "No Metal No Magic" Books Featuring

Get These Fun Elemental Periodic Table Activities at

MagicalPTElements.

Unicorn Periodic Table Bingo—Comes with 32 unique Bingo cards. Magical Elementals Bingo comes with 36.

Magical Elemental Game Cards—Makes great prizes. Fun to trade, too.

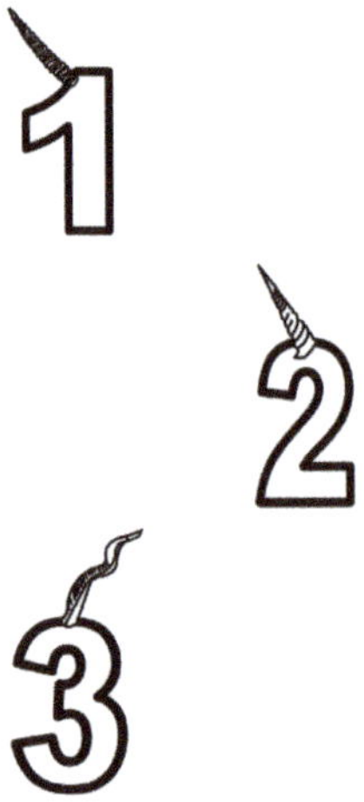

plus

1
2
3

Unicorn Horn

A
B
C

Alphabet

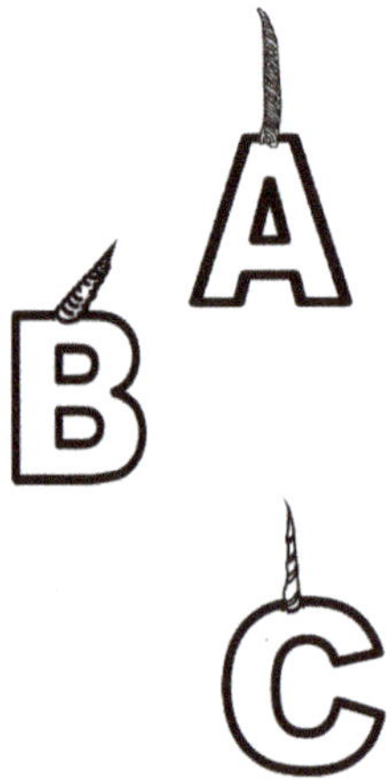

CLIP ART FOR YOUR GRAPHICS

Also browse activities at
https://www.magicalPTelements.com
for all kinds of printable downloads to make learning fun.

Printable Magical Elemental Activity Downloads

Fun Way For Students To Learn The Elements
Of The Periodic Table

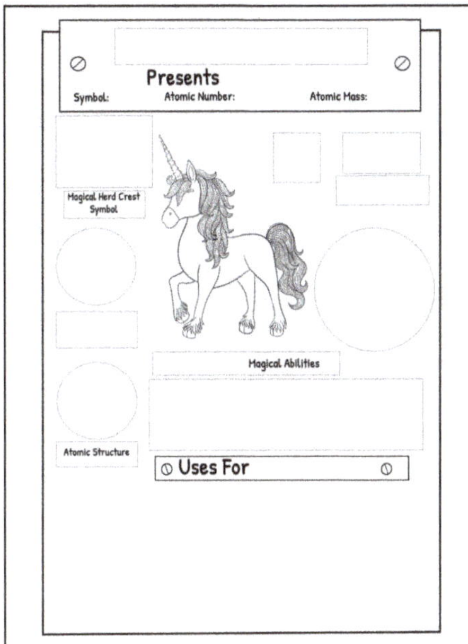

Blank Unicorn Element Card

Sample Unicorn Element Card

Magical Unicorn Elemental Research Sheet

Blank Research Sheet

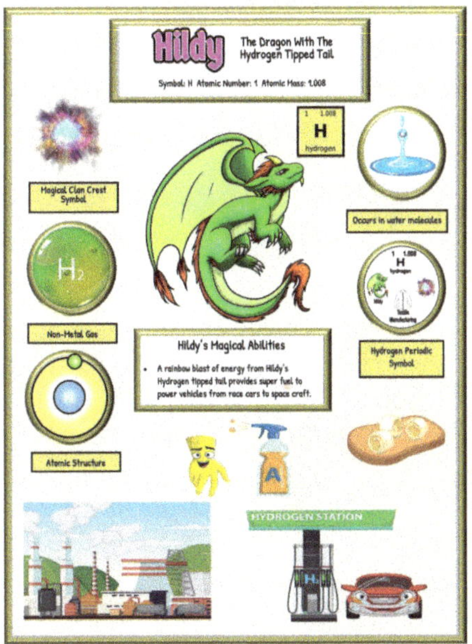

Sample Dragon Element Card

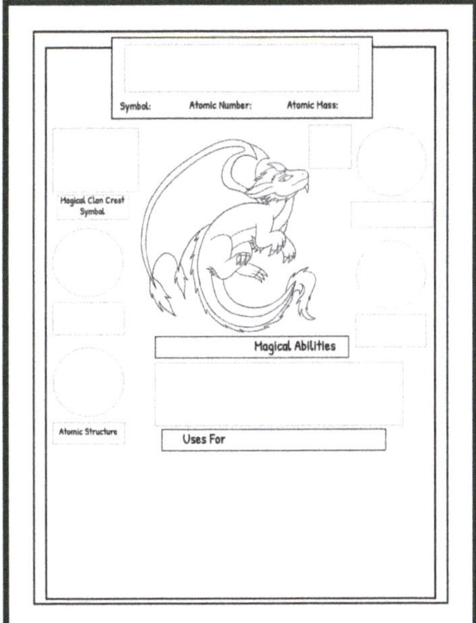

Blank Dragon Element Card

Magical Dragon Elemental Research Sheet

Blank Research Sheet

Using the sample Magical Elemental cards provided, have students select an element from the Periodic Table and a Magical Elemental Card Blank to create their own Magical Elemental Card. The blank and sample cards do not have to match.

You will receive a pdf containing either 26 unicorn or 26 dragon sample cards and blanks to be printed on 8 1/2 x 11 sized paper or card stock. The pdf also contains a Magical Elemental Research Sheet for the students to work on before creating their unique Periodic Table Elemental. They will also write a short paragraph describing their Unicorn or Dragon Elemental from that research.

Get These Fun Elemental Periodic Table Activity Sheets at *MagicalPTElements.com*

Don't Forget To Get A Tee Shirt

Featuring Your Favorite of the

118 Elements from the Periodic Table

Available in adult and kid sizes in many colors.

https://amzn.to/47NVZWN

This is Amerine's Tee Shirt Graphic.

Get it at https://www.amazon.com/dp/B0DMV14G7N

Dear Reader

I hope "No Metal No Magic Element 95 — Americium, Presented By Amerine, from The Magical Elements of the Periodic Table Book Series" with illustrations by Pranavva et al, has helped you learn some fun and interesting things about the magic of the element, Americium.

This is one of what will eventually be 118 books featuring periodic table elements presented by unicorns, dragons, wizards, knights and goblins. Keep checking regularly. Every one of the elements are amazing and very necessary to our everyday lives. All of the elements are

Techno-magical.

A lot of research went into every page of this book as well as the Magical Elements of the Periodic Table Books. There are just too many references to publish in this book but you can read and research them all at MagicalPTElements.com/MAUPT or /MDAPT or /MW1PT or / MW2PT or /MAKPT or /MRGPT. There, you can also access book related activity sheets and games to help make the learning process more fun.

Get ready made trading cards, lapel pins, tee shirts and more based on this book from Sybrina Publishing's No Metal No Magic Collection at Zazzle - **http://bit.ly/3km64Wg**

Would you like a 24" x 36" poster of the Elemental-Themed Periodic Table in this book? The best place to get it is at **https://bit.ly/49QMxBT** They have the sharpest images of any other poster printer around.

The Magical Elements of the Periodic Table books came into existence because of my Blue Unicorn—Journey To Osm books. If it weren't for their magical powers, based on the properties of the metals of their horns and hooves, I would have never come up with the idea to relate magical creatures to the periodic table. There's a metal horn unicorn story for every age group and they are all available at MagicalPTElements.com

If you enjoyed this book
please leave a nice review
at your favorite online book site.

No Metal No Magic

Song Lyrics

No metal, no Magic

No metal, no Magic

I can think of nothing more tragic

Than to have no metal or no magic

Metal makes everything magical.

Just ask a unicorn. . .

Preferably, one with a metal horn.

They'd say No metal, No magic.

Metal makes everything techno magical.

No metal, No magic

for two-leggers or unicorns.

No metal, No magic

Metal makes everything techno magical.

No metal, No magic

It's techno magical.

No metal, No magic

It might be very hard to believe but with

No metal, No magic

There'd be no technology.

No metal, No magic

Listen to this song at https://youtu.be/tcB8KDWAd8w

Watch the book trailer at https://youtu.be/NIX9fE7GJRI

Blue Unicorn

Ebooks, Audio and Print Books Available at all online book stores.

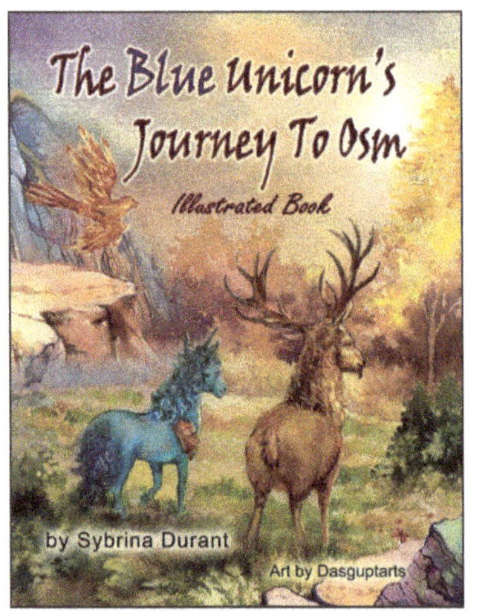

Illustrated
Book

Audio
Book

'Read & Color' Book
For Teens

Unicorn Periodic
Table Book

Fantasy
Novel

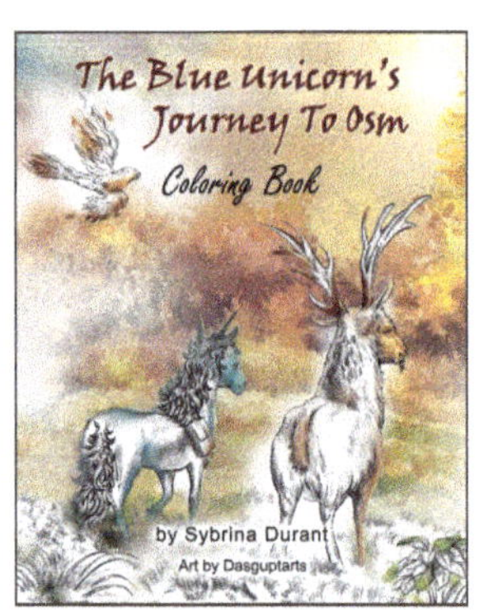

Coloring Book & Character
Introduction

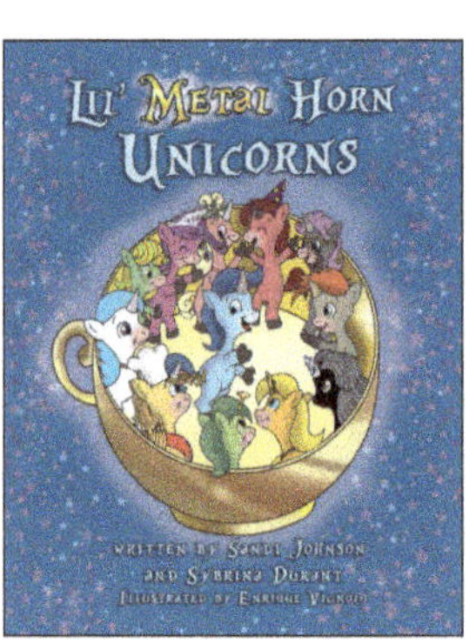

Picture Book For
Kids

Get these and more at

MagicalPTElements.com

and Sybrina.com

Journey To Osm—The Blue Unicorn's Tale

Back then, most places throughout MarBryn had wizards and sorcerers of some ability, or another. Most were trained in the ways of magical arts by the unicorns as part of their outreach program. Some two-leggers developed practical magical skills like making delicious feasts of tasty food appear out of thin air or purifying murky water around the land.

Others went through more extensive training to learn battle magic—like shooting powerful streams of energy from their swords.

Some were taught the art of holding the glow of the sun in magical globes, bringing light into the dark of night. These magical lights warded off the evil beings that were new and frightening products of dark magic.

With the rise of the sorcerer Magh, magical defense arts had become more important. The highest level of magical training involved sensing when others were in danger and learning to see into the future. Very few two-leggers ever reached that level because magic wasn't inherent in them the way it was for the unicorns. The metal of their horns and hooves were part of them as well as the very makeup of their blood, but two-leggers relied on learned magic via potions, charms, and incantations that required help from ingredients and forces more mystical in nature than any two-legger was ever born to be. Of course, controlling magic and projecting your intentions went far beyond merely following a recipe of sorts. It took being in touch with nature and the various elements to get the response a wizard desired. Much trial and error went into it, as well as faith and trust and the motives of the spell caster.

Magic was and is a practice that is never quite perfected even for the unicorns who must continue to hone and learn how to harness their powers. On rare occasion a wizard and unicorn had formed enough trust and a steadfast bond that prompted the unicorn to gift the wizard with a wand or staff embedded with the smallest sliver of metal from one of their hooves. This was rare but had happened and of course so had the desire for more power. Magh wasn't the first sorcerer with lust for more and throughout history there had been a handful of heinous acts against unicorns from those seeking their magic. Prior to Magh those wizards had failed to circumvent the protections nature had infused unicorn magic with so even after harvesting metal from their horns or hooves these sorcerers had gone mad trying to bend the will of nature and actually use their ill-gotten gains.

Magh, too, had gone mad or perhaps he'd already been so, but somehow he'd managed to harness the metal he harvested from his victims and continued to grow stronger rather than completely lose his mind like the others. How exactly remained a mystery to the unicorns and everyone else.

The wizards who'd honed their craft with the help and blessing of the unicorns tried to solve the riddle and even the best of them failed to uncover that secret. Some of MarBryn's natives took to magic naturally, while others struggled with the concept but no matter how adept they were. All of them fought valiantly against Magh's magic because with even a shred of knowledge they understood how the power shift would ultimately play out. They fought to the end but, in the end, only one sorcerer remained in MarBryn and now, Magh was in total control. But still, he was not satisfied. He wanted to control every living creature in the land.